Geology: A Comprehensive Introduction

Geology: A Comprehensive Introduction

Skylar Green

SYRAWOOD
PUBLISHING HOUSE
New York

Published by Syrawood Publishing House,
750 Third Avenue, 9th Floor,
New York, NY 10017, USA
www.syrawoodpublishinghouse.com

Geology: A Comprehensive Introduction
Skylar Green

International Standard Book Number: 978-1-64740-114-6 (Hardback)

Cataloging-in-Publication Data

Geology : a comprehensive introduction / Skylar Green.
p. cm.
Includes bibliographical references and index.
ISBN 978-1-64740-114-6
1. Geology. 2. Earth sciences. 3. Geology--Research. I. Green, Skylar.
QE26.3 .G46 2022
551--dc23

TABLE OF CONTENTS

PREFACE

It is with great pleasure that I present this book. It has been carefully written after numerous discussions with my peers and other practitioners of the field. I would like to take this opportunity to thank my family and friends who have been extremely supporting at every step in my life.

The branch of earth science which deals with the structure of the rocks and solid Earth along with the processes is known as geology. It involves the study of rock as it provides primary records of geologic history. There are various sub disciplines of geology, namely, biogeology, structural geology, stratigraphy, petrology, hydrogeology etc. A number of field, numerical and laboratory modeling methods are used to understand diverse processes related to Earth. Geological mapping is the use of geologic maps which are made to understand varied geological features. It comprises of structural mapping, surficial mapping and stratigraphic mapping. Varied dating methods are applied to provide information of geologic history and its timeline. This book provides comprehensive insights into the field of geology. Most of the topics introduced herein cover new techniques and the applications of this field. It will provide comprehensive knowledge to the readers.

The chapters below are organized to facilitate a comprehensive understanding of the subject:

Chapter - What is Geology?

The branch of earth science which deals with the solid Earth, the rocks of which it is comprised as well as the processes by which they change over time is referred to as geology. It also studies the structure of the Earth. This is an introductory chapter which will provide a brief introduction to the significant aspects of geology.

Chapter - Geologic Materials

Geologic materials refer to the various kinds of rock and unlithified material which are found on the surface of the Earth. The major types of rocks are igneous rocks, sedimentary rocks and metamorphic rocks. The topics elaborated in this chapter will help in gaining a better perspective about these geologic materials.

Chapter - Principles and Laws of Geology

There are various laws and principles which are studied within the discipline of geology. A few of them are the principle of original horizontality, principle of lateral continuity, principle of faunal succession, law of included fragments and law of superposition. This chapter discusses in detail these principles and laws of geology.

Chapter - Branches of Geology

Geology is a vast discipline which can be divided into numerous branches. A few of these are stratigraphy, structural geology, petrology, ethnogeology, historical geology and geochronology. This chapter has been carefully written to provide an easy understanding of these branches of geology.

Chapter - Geomorphology and its Processes

The scientific study of the origin and evolution of topographic and bathymetric features is referred to as geomorphology. The processes related to this field include aeolian processes, fluvial processes, weathering, metamorphism, etc. This chapter closely examines these processes of geomorphology to provide an extensive understanding of the subject.

Chapter - Field Methods in Geology

There are various field methods that are used in geology. Some of these are geologic mapping, radiometric dating and geophysical survey. The geological map is a special purpose map that is made to show features of Earth. The diverse applications of these field methods have been thoroughly discussed in this chapter.

Chapter - Applied Geology

The domain of geology which makes use of the principles and methods of pure geology to fulfill various material needs of humans is referred to applied geology. It encompasses diverse fields such as economic geology, engineering geology, environmental geology, forensic geology and geologic modeling. This chapter closely examines these subdomains of applied geology to provide an extensive understanding of the subject.

Skylar Green

What is Geology? 1

- **Geologic History of Earth**
- **Structure of the Earth**
- **Faults**
- **Folds**

The branch of earth science which deals with the solid Earth, the rocks of which it is comprised as well as the processes by which they change over time is referred to as geology. It also studies the structure of the Earth. This is an introductory chapter which will provide a brief introduction to the significant aspects of geology.

Geology deals with the study of the earth in which we live. The very fact that we live on the earth is enough reason to justify studying it. As we come to know about the earth, particularly its environment as well as its resources we understand and appreciate it. We realize that our earth is the most important body in the universe.

In a broader view however the earth perhaps has to compete with other celestial systems. It is known that there are stars having their own planets. Humans have not yet devised telescopes powerful enough to sight them. Yet we are convinced of their existence from the wavering motions of the stars. The earth is an average planet orbiting an average sized star, the sun.

The earth seems to be a medium sized planet but we however realize its uniqueness as it has abundant water exemplified by the oceans. The earth has an atmosphere which supports life.

The surface temperature is controlled mainly by the distance from the sun, and it makes various features possible and in turn these features make life possible on the earth. The development and history of life are pertinent aspects of historical geology.

We are now aware, thanks to the space programmes that the earth is unique in possessing a magnetic field and this magnetic field is known to be caused by the liquid iron core which is also known to have stored the energy that causes formation of various surface features like mountain ranges.

These features are susceptible to the processes of deformation and erosion of the earth's surface which are the chief aspects of physical geology. Geology by and large is the study of the various consequences of the unique features of the earth.

Geology has significantly contributed to the human civilization intellectually and economically. Understanding the age of the earth and the development of an absolute time scale are among the great concepts realized from geological studies. It is worthy to note that geology differs from most other subjects in that it is concerned with absolute time.

Time as we often deal with in various equations of physics and chemistry is mostly concerned with rates of change and such time gauging is relative – not absolute. Geologic time extends back to even 5 billion years or more, to when earth formed.

Geology is concerned with enormous periods of time when gauged against human experience. Yet another aspect brought to light from geology is that continuous change both biological and physical which has been taking place on the earth.

The economic contributions of geology to human civilization show in many ways that it is a useful practical subject. Geologic knowledge helps to locate and mine mineral resources. With the exception of water and soil, all mineral resources such as sand and gravel, petroleum, coal and metals are non-renewable. Once they are mined, they are gone and we have to search for new deposits.

Geologists have been able to discover the deposits of metals and energy producing minerals on which our civilization is based. Geologists are also concerned with other economic problems such as urbanization. As large cities are developed huge structures like tall and massive buildings, bridges and dams are constructed and a knowledge of geology is very helpful in designing the foundations of these structures.

There are instances of large as well as small structures that have failed through neglect of geologic principles. Dams fail if they are built close to active faults or on porous foundations.

There are instances of highways newly constructed which have been washed out or blocked by landslides during the very first rainy season after their construction. Some homes built on hill sides have been destroyed by mud flows and landslides. Geologists have recognized the need for earthquake resistant structures and have helped in their design.

Geology is based on observations and tries to find out the history of the earth by logical observations and applying also other branches of science like physics, chemistry and biology. For instance, using fossils to date or establish contemporaneity of rock strata is based on the simple basic principle that life has changed during the long history of the earth.

This principle has been arrived at as a result of careful observations made at many places. The scale and range of most problems cannot be managed by laboratory study. For instance, it is not possible to bring a volcano into a laboratory even though some facets of volcanoes can be studied indoors. It is not also possible to simulate geologic time in any experiment.

The geologist adopts the method of multiple working hypotheses to verify his theories and to attempt to reach the best reasoned theory. This process needs to process as many hypotheses as possible and also the skill to devise ways to test each theory. It may not always be possible to reach a unique solution.

In this subject we study by considering the crust – the outer few kilometres – of the earth. We study the minerals and rocks of the crust since they bring to light about the composition of the crust and the various processes that take place within the crust as well as on its surface. We study erosion, the most dominant surface process.

We study the interactions of the processes in the crust and deeper parts of the earth and learn how the present surface features of the earth took place. We study how the river moves its load of sediment or how petroleum moves beneath the ground.

We study the mechanisms involved in earthquakes. An important aim of geology is to develop an understanding of the earth materials and how these materials undergo changes and modifications through the action of natural agents and forces over a period of time.

We realize that the various features of the earth were not formed at the same time. We try to answer many basic questions through the study of geology. We try to find out how many things happened.

We try to understand phenomena like, the growth of a volcano or the movement of a glacier or the development of sinuous river channels or how rocks undergo distortions due to external forces and so on. We study the formation of the mountain ranges or how the coal fields were formed.

In short, geological changes are vast and they happened and continue to happen. The earth is continuously being shaped and reshaped by the slow geological changes in its features.

Role of Geologists Geologists are involved in a variety of fields. They work with minerals, rocks, ores and fossils.

Petroleum geologists seek deposits of petroleum and natural gas which constitute our main natural fuels and our most precious mineral product.

Mining geologists are involved in searching for ore deposits of iron, uranium, lead and zinc and also non-metallic deposits like coal, limestone, salt, gypsum, potash. Groundwater geologists help in locating supplies of underground water for domestic and industrial use.

Petrologists make a study of composition and origin of various rocks. Paleontologists are involved in the study of fossils, the remains or traces of past life which are naturally preserved in rocks. Fossils indicate an orderly development from the most primitive forms in very old rocks to the remnants of life of modern times. Study of fossils in well cores in drilling for oil can be used to identify the relative ages of rocks.

Stratigraphers make a study of the order in which the rocks were formed so as to guide the search for petroleum, uranium and such materials. Structural geologists make a study of the architecture or arrangement of the rocks of the earth's crust.

Geomorphologists study the nature and origin of landscapes. Engineering geologists devote themselves to the application of geology to such projects as dam sites, tunnels, water reservoirs, highway and railroad construction and construction materials. Geologists receive assistance from exploration geophysicists, seismologists, geochemists and such specialists as paleobotanists and offer assistance to engineers, miners, soil conservationists etc.

Importance of Geology in Civil Engineering Practice

A civil engineer has to deal with a variety of problems where a knowledge of the subject of geology will be very useful.

Some important fields of application of the knowledge of geology are the following:

1. Geology provides a systematic knowledge of materials, their occurrence and their properties. The sources of materials, the various types of materials and their characteristics are all geological fundamentals.

2. Problems we come across in foundations are directly geological. Structures, whether buildings bridges, highways, railroads, dams etc. are all built on or in some natural materials.

3. Knowledge of geology will be useful in open or underground excavations.

4. Knowledge of groundwater occurrence and groundwater hydrology are useful while dealing with various constructions and engineering fields like sanitary engineering, land drainage, water supply, control of landslide etc.

5. Knowledge of surface waters helps us to understand the processes of erosion, transportation and deposition. It is also needed to act on river control, coastal and harbour works, soil conservation etc.

6. Geological reports, geological and topographical maps help in planning various projects.

Various Branches of Geology

For the study of geology is divided into the following branches:

1. Physical geology deals with the principles and the various processes that are acting on the surface of the earth modifying its relief features over a period of time. This subject deals with the action of wind, rivers, glaciers, volcanoes, earthquakes, landslides etc. in bringing about changes in the surface features of the earth.

2. Mineralogy deals with the various minerals which form the building units of rocks which constitute the earth's crust. We study in this subject the formation, occurrence and the properties of the minerals.

3. Petrology deals with various types of rocks forming the earth's crust. We study in this subject about the formation of the different types of rocks.

4. Structural Geology deals with the arrangement of the structured relations of rocks. This subject deals with how folds, faults and joints occur in rocks.

5. Palaeontology deals with the study of the remains of ancient life, that are found in rocks.

6. Stratigraphy deals with the history of the earth as depicted by its rocks and particularly the record of events revealed in the rocks.

7. Economic geology deals with the occurrence, origin and distribution of earth materials

which are valuable to humanity. This subject deals with the application of these materials in industry and engineering fields.

Processes of Geological Change

Gradation Rocks

Gradation Rocks which are exposed to the surface of the earth are continuously subjected to gradation. Gradation may be either degradation (wearing down by water, wind, ice) on aggradation building up of rock formations by deposition of the materials degraded.

Degradation generally takes place above the sea level while aggradation occurs below the sea level. During rain falls, the rain water fills creeks and rivers and reaches lakes and seas.

The water carries with it rock particles and drops them in lakes and along rivers and in sea where they become sand beds and mud beds. Sea waves beat continuously beat the coasts wearing away the land.

Winds carry rock particles or roll them on the surface of the earth and deposit them to become sand dunes or layers and films of dust. Glaciers (ice in motion) carry rock fragments and as ice melts, these are deposited. The process of gradation is slow but continuous.

Diastrophism

Diastrophism refers to the movement of solid parts of the earth with respect to each other. At many places rock beds get raised or lowered. Large areas of the earth's surface have been submerged below the sea level or elevated above the sea level.

Diastrophism generally determines the nature of gradation. Degradation is dominant while aggradation is dominant below the sea level. Degradation breaks down the rocks and transports their particles to the sea where new beds are formed.

By diastrophism these beds are raised above the sea and exposed again to degradation. When the sediments are deposited in the sea, they are nearly flat-lying. If they are found to be tilted, it gives the indication that they have been disturbed by diastrophic movements.

Vulcanism

Vulcanism relates to all phenomena concerning movement of molten rocks and the formation of their products. Large parts of the earth's surface are made up of rocks which have solidified from the molten state. These are igneous rocks. Some igneous rocks are extrusive rocks which are formed by solidification of ejected molten and semi molten rock ejected out of the throats of volcanoes.

If the molten rocks solidify at some depths below the earth's surface the rocks formed are called intrusive rocks. Over a period of time these intrusive rocks may get exposed at the surface due to the removal of material above them in the process of erosion. The igneous rocks so exposed are attacked by weathering and geological agents and are broken down and carried by air stream, glaciers etc. and deposited to form sediments.

Geologic History of Earth

Geologic history of Earth refers to the evolution of the continents, oceans, atmosphere, and biosphere. The layers of rock at Earth's surface contain evidence of the evolutionary processes undergone by these components of the terrestrial environment during the times at which each layer was formed. By studying this rock record from the very beginning, it is thus possible to trace their development and the resultant changes through time.

Pregeologic Period

From the point at which the planet first began to form, the history of Earth spans approximately 4.6 billion years. The oldest known rocks—the faux amphibolites of the Nuvvuagittuq greenstone belt in Quebec, Canada—however, have an isotopic age of 4.28 billion years. There is in effect a stretch of approximately 300 million years for which no geologic record for rocks exists, and the evolution of this pregeologic period of time is, not surprisingly, the subject of much speculation. To understand this little-known period, the following factors have to be considered: the age of formation at 4.6 billion years ago, the processes in operation until 4.3 billion years ago, the bombardment of Earth by meteorites, and the earliest zircon crystals.

It is widely accepted by both geologists and astronomers that Earth is roughly 4.6 billion years old. This age has been obtained from the isotopic analysis of many meteorites as well as of soil and rock samples from the Moon by such dating methods as rubidium–strontium and uranium–lead. It is taken to be the time when these bodies formed and, by inference, the time at which a significant part of the solar system developed. When the evolution of the isotopes of lead-207 and lead-206 is studied from several lead deposits of different age on Earth, including oceanic sediments that represent a homogenized sample of Earth's lead, the growth curve of terrestrial lead can be calculated, and, when this is extrapolated back in time, it is found to coincide with the age of about 4.6 billion years measured on lead isotopes in meteorites. Earth and the meteorites thus have had similar lead isotope histories, and so it is concluded that over a period of about 30 million years they condensed or accreted as solid bodies from a primeval cloud of interstellar gas and dust—the so-called solar nebula from which the entire solar system is thought to have formed—at about the same time.

Models developed from the comparison of lead isotopes in meteorites and the decay of hafnium-182 to tungsten-182 in Earth's mantle, however, suggest that approximately 100 million years elapsed between the beginning of the solar system and the conclusion of the accretion process that formed Earth. These models place Earth's age at approximately 4.5 billion years old.

Particles in the solar nebula condensed to form solid grains, and with increasing electrostatic and gravitational influences they eventually clumped together into fragments or chunks of rock. One of these planetesimals developed into Earth. The constituent metallic elements sank toward the centre of the mass, while lighter elements rose toward the top. The lightest ones (such as hydrogen and helium) that might have formed the first, or primordial, atmosphere probably escaped into outer space. In these earliest stages of terrestrial accretion heat was generated by three possible phenomena: (1) the decay of short-lived radioactive isotopes, (2) the gravitational energy released from the sinking of metals, or (3) the impact of small planetary bodies (or planetesimals). The increase in temperature became sufficient to heat the entire planet. Melting at depth produced liquids that were gravitationally light and thus rose toward the surface and crystallized to form the earliest crust. Meanwhile, heavier liquids rich in iron, nickel, and perhaps sulfur separated out and sank under gravity, giving rise to the core at the centre of the growing planet; and the lightest volatile elements were able to rise and escape by outgassing, which may have been associated with surface volcanic activity, to form the secondary atmosphere and the oceans. This chemical process of melting, separation of material, and outgassing is referred to as the differentiation of the Earth. The earliest thin crust was probably unstable and so foundered and collapsed to depth. This in turn generated more gravitational energy, which enabled a thicker, more stable, longer-lasting crust to form. Once Earth's interior (or its mantle) was hot and liquid, it would have been subjected to large-scale convection, which may have enabled oceanic crust to develop above upwelling regions. Rapid recycling of crust–mantle material occurred in convection cells, and in this way the earliest terrestrial continents may have evolved during the 300-million-year gap between the formation of Earth and the beginning of the rock record. It is known from direct observation that the surface of the Moon is covered with a multitude of meteorite craters. There are about 40 large basins attributable to meteorite impact. Known as maria, these depressions were filled in with basaltic lavas caused by the impact-induced melting of the lunar mantle. Many of these basalts have been analyzed isotopically and found to have crystallization ages of 3.9 to 4 billion years. Research has shown that Earth, with a greater attractive mass than the Moon, must have undergone more extensive meteorite bombardment. According to the English-born geologist Joseph V. Smith, a minimum of 500 to 1,000 impact basins were formed on Earth within a period of about 100 to 200 million years prior to 3.95 billion years ago. Moreover, plausible calculations suggest that this estimate represents merely the tail end of an interval of declining meteorite bombardment and that about 20 times as many basins were formed in the preceding 300 million years. Such intense bombardment would have covered most of Earth's surface, with the impacts causing considerable destruction of the terrestrial crust up to 4 billion years ago.

An exciting discovery was made in 1983 by William Compston and his research group at the Australian National University with the aid of an ion microprobe. Compston and his associates found that a water-laid clastic sedimentary quartzite from Mount Narryer in western Australia contained detrital zircon grains that were 4.18 billion years old. In 1986 they further discovered that one zircon in a conglomerate only 60 km (about 37 miles) away was 4.276 billion years old; 16 other grains were determined to be the same age or slightly younger. In 2014 American geochemist John Valley and colleagues discharged lead atoms at one of the zircon crystals discovered in western Australia's Jack Hills and discovered that the crystals are more than 4.4 billion years old. This mineral is the oldest dated material on Earth. The rocks from which the zircons in the quartzites and conglomerates were derived have either disappeared or have not yet been found. The ages of these single zircon grains are at least roughly 100 million years older than those of the oldest known intact rocks.

Development of the Atmosphere and Oceans

Formation of the Secondary Atmosphere

Earth's secondary atmosphere began to develop at the time of planetary differentiation, probably in connection with volcanic activity. Its component gases, however, were most likely very different from those emitted by modern volcanoes. Accordingly, the composition of the early secondary atmosphere was quite distinct from that of today's atmosphere. Carbon monoxide, carbon dioxide, water vapour, and methane predominated; however, free oxygen could not have been present, since even modern volcanic gases contain no oxygen. It is therefore assumed that the secondary atmosphere during the Archean Eon (4 billion to 2.5 billion years ago) was anoxygenic. The free oxygen that makes up the bulk of the present atmosphere evolved over geologic time by two possible processes. First, solar ultraviolet radiation (the short-wavelength component of sunlight) would have provided the energy needed to break up water vapour into hydrogen, which escaped into space, and free oxygen, which remained in the atmosphere. This process was in all likelihood important before the appearance of the oldest extant rocks, but after that time the second process, organic photosynthesis, became predominant. Primitive organisms, such as blue-green algae (or cyanobacteria), cause carbon dioxide and water to react by photosynthesis to produce carbohydrates, which they need for growth, repair, and other vital functions, and this reaction releases free oxygen. The discovery of stromatolites (layered or conical sedimentary structures formed by sediment-binding marine algae) in 3.5-billion-year-old limestones in several parts of the world indicates that blue-green algae existed by that time. The presence of such early carbonate sediments is evidence that carbon dioxide was present in the atmosphere, and it has been calculated that it was at least 100 times greater than the amount in the present-day atmosphere. It can be assumed that such abundant carbon dioxide would have caused retention of heat, resulting in a greenhouse effect and a hot atmosphere.

What happened to all the oxygen that was released? It might be surprising to learn that it took at least 1 billion years before there was sufficient oxygen in the atmosphere for oxidative diagenesis to give rise to red beds (sandstones that are predominantly red in colour due to fully oxidized iron coating individual grains) and that 2.2 billion years passed before a large number of life-forms could evolve. An idea formulated by the American paleontologist Preston Cloud has been widely accepted as an answer to this question. The earliest primitive organisms produced free oxygen as a by-product, and in the absence of oxygen-mediating enzymes it was harmful to their living cells and had to be removed. Fortunately for the development of life on the early Earth there was extensive volcanic activity, which resulted in the deposition of much lava, the erosion of which released enormous quantities of iron into the oceans. This ferrous iron is water-soluble and therefore could be easily transported, but it had to be converted to ferric iron, which is highly insoluble, before it could be precipitated as iron formations. In short, the organisms produced the oxygen and the iron formations accepted it. Iron formations can be found in the earliest sediments (those deposited 3.8 billion years ago) at Isua in West Greenland, and thus this process must have been operative by this time. Iron formations dating to early Precambrian time (4.6 billion to 541 million years ago) are so thick and common that they provide the major source of the world's iron. Large quantities of iron continued to be deposited until about 2 billion years ago, after which time the formations decreased and disappeared from the sedimentary record. Sulfides also accepted oxygen in the early oceans to be deposited as sulfates in evaporites, but such rocks are easily destroyed. One finds, nonetheless, 3.5-billion-year-old barite/gypsum-bearing evaporites up to 15 metres (about 49 feet) thick and at least 25 km (15.5 miles) in extent in the Pilbara region of Western Australia. It seems likely that the excess iron in the early oceans was finally cleared out by about 1.7 billion years ago, and this decrease in the deposition of iron formations resulted in an appreciable rise in the oxygen content of the atmosphere, which in turn enabled more eolian red beds to form. Further evidence of the lack of oxygen in the early atmosphere is provided by detrital uraninite and pyrite and by paleosols—i.e., fossil soils. Detrital uraninite and pyrite are readily oxidized in the presence of oxygen and thus do not survive weathering processes during erosion, transport, and deposition in an oxygenous atmosphere. Yet, these minerals are well preserved in their original unoxidized state in conglomerates that have been dated to be more than 2.2 billion years old on several continents. Paleosols also provide valuable clues, as they were in equilibrium with the prevailing atmosphere. From analyses of early Precambrian paleosols it has been determined that the oxygen content of the atmosphere 2.2 billion years ago was one hundredth of the present atmospheric level (PAL).

Fossils of eukaryotes, which are organisms that require an oxygen content of about 0.02 PAL, bear witness to the beginning of oxidative metabolism. The first microscopic eukaryotes appeared about 1.4 billion years ago. Life-forms with soft parts, such as jellyfish and worms, developed in profusion, albeit locally, toward the end of the Precambrian about 650 million years ago, and it is estimated that this corresponds to an oxygen level of 0.1 PAL. By the time land plants first appeared, roughly 400 million years ago, atmospheric oxygen levels had reached their present values.

Development of the Oceans

Volcanic degassing of volatiles, including water vapour, occurred during the early stages of crustal formation and gave rise to the atmosphere. When the surface of Earth had cooled to below 100 °C (212 °F), the hot water vapour in the atmosphere would have condensed to form the early oceans. The existence of 3.5-billion-year-old stromatolites is, as noted above, evidence of the activity of blue-green algae, and this fact indicates that Earth's surface must have cooled to below 100 °C by this time. Also, the presence of pillow structures in basalts of this age attests to the fact that these lavas were extruded under water, and this probably occurred around volcanic islands in the early ocean. The abundance of volcanic rocks of Archean age is indicative of the continuing role of intense volcanic degassing, but since early in the Proterozoic Eon (2.5 billion to 541 million years ago) much less volcanic activity has occurred. Until about 2 billion years ago there was substantial deposition of iron formations, cherts, and various other chemical sediments, but from roughly that time onward the relative proportions of different types of sedimentary rock and their mineralogy and trace element compositions have been very similar to their Phanerozoic equivalents (that is, rocks laid down during the Phanerozoic Eon [541 million years ago to the present]); it can be inferred from this relationship that the oceans achieved their modern chemical characteristics and sedimentation patterns from approximately 2 billion years ago. By the late Precambrian, some 1 billion years ago, ferric oxides were chemically precipitated, indicating the availability of free oxygen. During Phanerozoic time, the oceans have been steady-state chemical systems, continuously reacting with the minerals added to them via drainage from the continents and with volcanic gases at the oceanic ridges.

Time Scales

The geologic history of Earth covers more than 4.5 billion years of time. Different types of phenomena and events in widely separated parts of the world have been correlated using an internationally acceptable, standardized time scale. There are, in fact, two geologic time scales. One is relative, or chronostratigraphic, and the other is absolute, or chronometric. The chronostratigraphic scale has evolved since the mid-1800s and concerns the relative order of strata. Important events in its development were the realization by English engineer and geologist William Smith that in a horizontal sequence of sedimentary strata what is now an upper stratum was originally deposited on a lower one and the discovery by Scottish geologist James Hutton that an unconformity (discontinuity) indicates a significant gap in time. Furthermore, the presence of fossils throughout Phanerozoic sediments has enabled paleontologists to construct a relative order of strata. As was explained earlier, at specific stratigraphic boundaries certain types of fossils either appear or disappear or both in some cases. Such biostratigraphic boundaries separate larger or smaller units of time that are defined as eons, eras, periods, epochs, and ages.

The chronometric scale is of more recent origin. It was made possible by the development of mass spectrometers during the 1920s and their use in geochronological laboratories for radiometric dating. The chronometric scale is based on specific units of duration and on the numerical ages that are assigned to the aforementioned chronostratigraphic boundaries. The methods used entail the isotopic analyses of whole rocks and minerals of element pairs, such as potassium–argon, rubidium–strontium, uranium–lead, and samarium–neodymium. Another radiometric time scale has been developed from the study of the magnetization of basaltic lavas of the ocean floor. As such lavas were extruded from the mid-oceanic ridges, they were alternately magnetized parallel and opposite to the present magnetic field of Earth and are thus referred to as normal and reversed. A magnetic-polarity time scale for the stratigraphy of normal and reversed magnetic stripes can be constructed back as far as 280–260 million years ago, which is the age of the oldest extant segment of ocean floor.

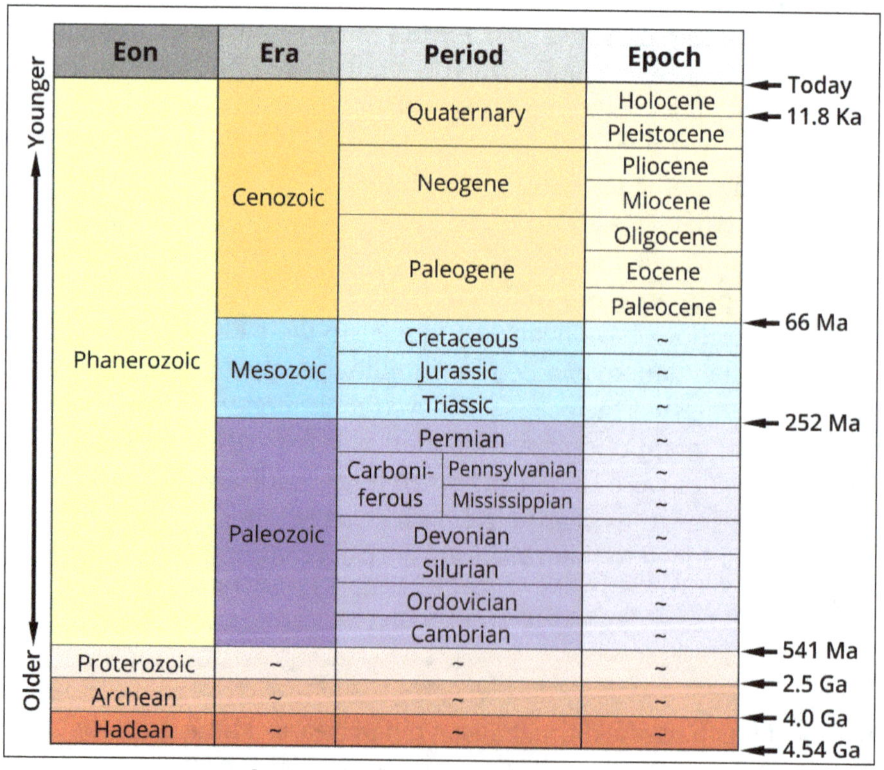

The stratigraphic chart of geologic time.

Structure of the Earth

The internal structure of the Earth is layered in spherical shells: an outer silicate solid crust, a highly viscous asthenosphere and mantle, a liquid outer corethat is much less viscous than the mantle, and a solid inner core. Scientific understanding of the internal structure of the Earth is based on observations of topography and bathymetry, observations of rock in outcrop, samples brought to the surface from greater depths by volcanoes or volcanic activity, analysis of the seismic waves that pass through the Earth, measurements of the gravitational and magnetic fieldsof the Earth, and experiments with crystalline solids at pressures and temperatures characteristic of the Earth's deep interior.

Mass

The force exerted by Earth's gravity can be used to calculate its mass. Astronomers can also calculate Earth's mass by observing the motion of orbiting satellites. Earth's average density can be determined through gravimetric experiments, which have historically involved pendulums.

The mass of Earth is about 6×10^{24} kg.

Structure

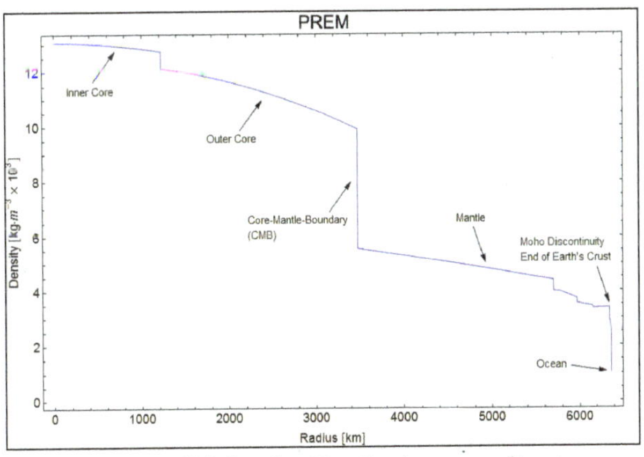

Earth's radial density distribution according to the preliminary reference earth model (PREM).

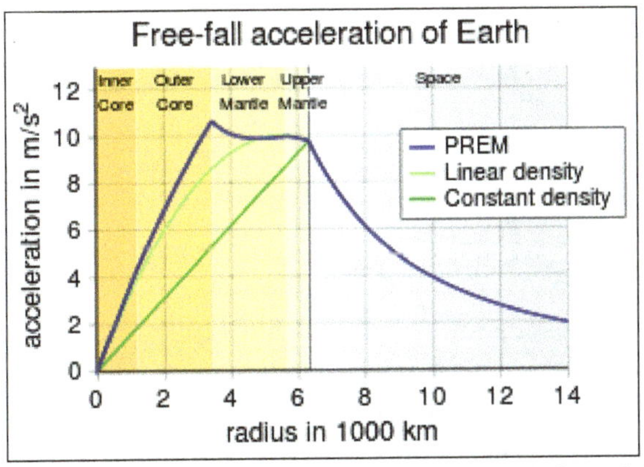

Earth's gravity according to the preliminary reference earth model (PREM). Comparison to approximations using constant and linear density for Earth's interior.

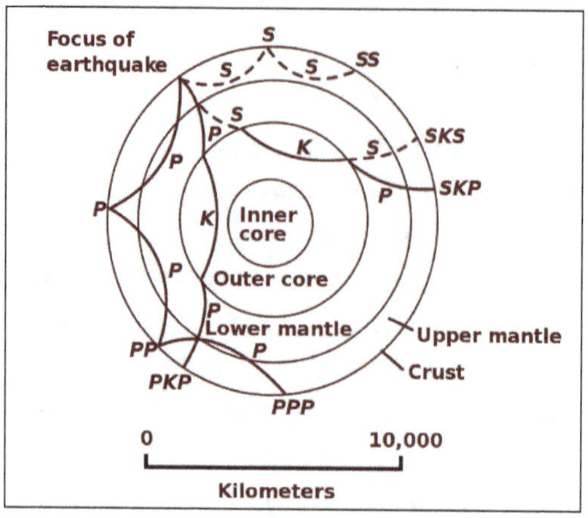

Mapping the interior of Earth with earthquake waves.

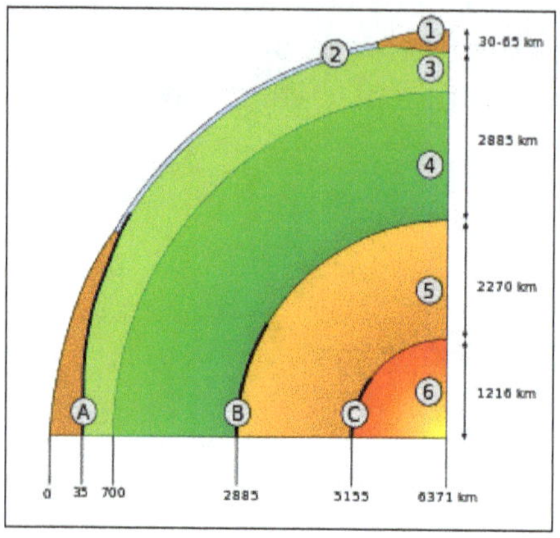

Schematic view of the interior of Earth's. 1. continental crust – 2. oceanic crust – 3. upper mantle – 4. lower mantle – 5. outer core – 6. inner core – A: Mohorovičić discontinuity – B: Gutenberg Discontinuity – C: Lehmann–Bullen discontinuity.

The structure of Earth can be defined in two ways: by mechanical properties such as rheology, or chemically. Mechanically, it can be divided into lithosphere, asthenosphere, mesospheric mantle, outer core, and the inner core. Chemically, Earth can be divided into the crust, upper mantle, lower mantle, outer core, and inner core. The geologic component layers of Earth are at the following depths below the surface:

Depth (km)	Layer
0–80	Lithosphere (locally varies between 5 and 200 km)
0–35	Crust (locally varies between 5 and 70 km)
35–2,890	Mantle
80–220	Asthenosphere
410–660	Transition zone
35–660	Upper mantle
660–2,890	Lower mantle
2,740–2,890	D″ layer
2,890–5,150	Outer core
5,150–6,360	Inner core

The layering of Earth has been inferred indirectly using the time of travel of refracted and reflected seismic waves created by earthquakes. The core does not allow shear waves to pass through it, while the speed of travel (seismic velocity) is different in other layers. The changes in seismic velocity between different layers causes refraction owing to Snell's law, like light bending as it passes through a prism. Likewise, reflections are caused by a large increase in seismic velocity and are similar to light reflecting from a mirror.

Crust

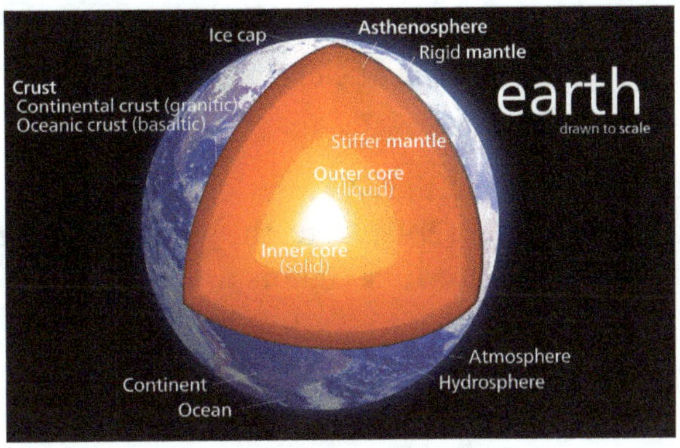

The internal structure of Earth.

In geology, the crust is the outermost solid shell of a rocky planet, dwarf planet, or natural satellite. It is usually distinguished from the underlying mantle by its chemical makeup; however, in the case of icy satellites, it may be distinguished based on its phase (solid crust vs. liquid mantle).

The crusts of Earth, Moon, Mercury, Venus, Mars, Io, and other planetary bodies formed via igneous processes, and were later modified by erosion, impact cratering, volcanism, and sedimentation.

Most terrestrial planets have fairly uniform crusts. Earth, however, has two distinct types: continental crust and oceanic crust. These two types have different chemical compositions and physical properties, and were formed by different geological processes.

Types of Crust

Planetary geologists divide crust into three categories, based on how and when they formed.

Primary Crust/Primordial Crust

This is a planet's "original" crust. It forms from solidification of a magma ocean. Toward the end of planetary accretion, the terrestrial planets likely had surfaces that were magma oceans. As these cooled, they solidified into crust. This crust was likely destroyed by large impacts and re-formed many times as the Era of Heavy Bombardment drew to a close.

The nature of primary crust is still debated: its chemical, mineralogic, and physical properties are unknown, as are the igneous mechanisms that formed them. This is because it is difficult to study: none of Earth's primary crust has survived to today. Earth's high rates of erosion and crustal recycling from plate tectonics has destroyed all rocks older than about 4 billion years, including whatever primary crust Earth once had.

However, geologists can glean information about primary crust by studying it on other terrestrial planets. Mercury's highlands might represent primary crust, though this is debated. The anorthosite highlands of the Moon are primary crust, formed as plagioclase crystallized out of the Moon's initial magma ocean and floated to the top; however, it is unlikely that Earth followed a similar pattern, as the Moon was a water-less system and Earth had water. The Martian meteorite ALH84001 might represent primary crust of Mars; however, again, this is debated. Like Earth, Venus lacks primary crust, as the entire planet has been repeatedly resurfaced and modified.

Secondary Crust

Secondary crust is formed by partial melting of silicate materials in the mantle, and so is usually basaltic in composition.

This is the most common type of crust in the Solar System. Most of the surfaces of Mercury, Venus, Earth, and Mars comprise secondary crust, as do the lunar maria. On Earth, we see secondary crust forming primarily at mid-ocean spreading centers, where the adiabatic rise of mantle causes partial melting.

Tertiary Crust

Tertiary crust is more chemically-modified than either primary or secondary. It can form in several ways:

- Igneous processes: Partial-melting of secondary crust, coupled with differentiation or dehydration.
- Erosion and sedimentation: Sediments derived from primary, secondary, or tertiary crust.

The only known example of tertiary crust is the continental crust of the Earth. It is unknown whether other terrestrial planets can be said to have tertiary crust, though the evidence so far suggests that they do not. This is likely because plate tectonics is needed to create tertiary crust, and Earth is the only planet in our Solar System with plate tectonics.

Earth's Crust

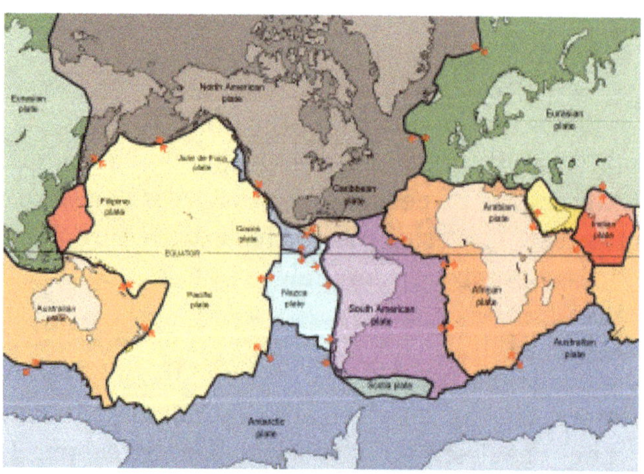

Plates in the crust of Earth.

Structure

The crust is a thin shell on the outside of the Earth, accounting for less than 1% of Earth's volume. It is the top component of lithosphere: a division of Earth's layers that includes the crust and the upper part of the mantle. The lithosphere is broken into tectonic plates that move, allowing heat to escape from the interior of the Earth into space.

The crust lies on top of the mantle, a configuration that is stable because the upper mantle is made of peridotite and so is significantly denser than the crust. The boundary between the crust and mantle is conventionally placed at the Mohorovičić discontinuity, a boundary defined by a contrast in seismic velocity.

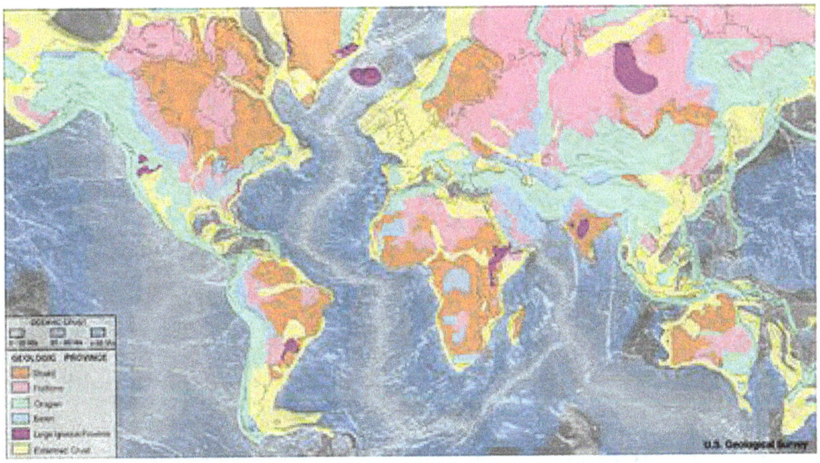

Geologic provinces of the world (USGS).

	Shield		Oceanic crust:
	Platform		Extended crust
	Orogen		0–20 Ma
	Basin		20–65 Ma
	Large igneous province		>65 Ma

The crust of the Earth is of two distinctive types:

1. Oceanic: 5 km (3 mi) to 10 km (6 mi) thick and composed primarily of denser, more maficrocks, such as basalt, diabase, and gabbro.

2. Continental: 30 km (20 mi) to 50 km (30 mi) thick and mostly composed of less dense, more felsic rocks, such as granite.

Because both continental and oceanic crust are less dense than the mantle below, both types of crust "float" on the mantle. This is isostasy, and it's also one of the reasons continental crust is higher than oceanic: continental is less dense and so "floats" higher. As a result, water pools in above the oceanic crust, forming the oceans.

The temperature of the crust increases with depth, reaching values typically in the range from about 200 °C (392 °F) to 400 °C (752 °F) at the boundary with the underlying mantle. The temperature increases by as much as 30 °C (54 °F) for every kilometer locally in the upper part of the crust, but the geothermal gradient is smaller in deeper crust.

Composition

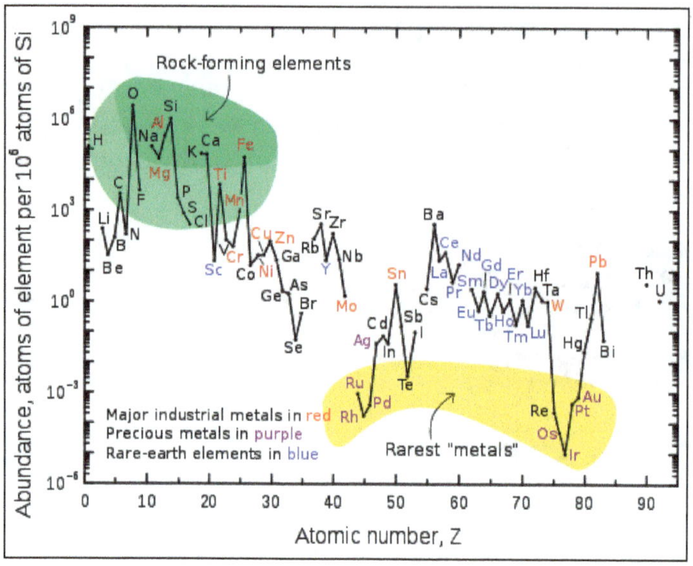

Abundance (atom fraction) of the chemical elements in Earth's upper continental crust as a function of atomic number. The rarest elements in the crust (shown in yellow) are not the heaviest, but are rather the siderophile (iron-loving) elements in the Goldschmidt classification of elements. These have been depleted by being relocated deeper into Earth's core. Their abundance in meteoroid materials is higher. Additionally, tellurium and selenium have been depleted from the crust due to formation of volatile hydrides.

The continental crust has an average composition similar to that of andesite. The most abundant minerals in Earth's continental crustare feldspars, which make up about 41% of the crust by weight, followed by quartz at 12%, and pyroxenes at 11%. Continental crust is enriched in incompatible elementscompared to the basalticocean crust and much enriched compared to the underlying mantle. Although the continental crust comprises only about 0.6 weight percent of the silicate on Earth, it contains 20% to 70% of the incompatible elements.

Most Abundant Elements of Earth's Crust	Approximate % by weight
O	46.6
Si	27.7
Al	8.1
Fe	5.0
Ca	3.6
Na	2.8

K	2.6
Mg	1.5

Oxide	Percent
SiO_2	60.6
$Al2O_3$	15.9
CaO	6.4
MgO	4.7
Na_2O	3.1
Fe as FeO	6.7
K_2O	1.8
TiO_2	0.7
P_2O_5	0.1

All the other constituents except water occur only in very small quantities and total less than 1%. Estimates of average density for the upper crust range between 2.69 and 2.74 g/cm³ and for lower crust between 3.0 and 3.25 g/cm³.

Formation and Evolution

Earth formed approximately 4.6 billion years ago from a disk of dust and gas orbiting the newly formed Sun. It formed via accretion, where planetesimals and other smaller rocky bodies collided and stuck, gradually growing into a planet. This process generated an enormous amount of heat, which caused early Earth to melt completely. As planetary accretion slowed, Earth began to cool, forming its first crust, called a primary or primordial crust. This crust was likely repeatedly destroyed by large impacts, then reformed from the magma ocean left by the impact. None of Earth's primary crust has survived to today; all was destroyed by erosion, impacts, and plate tectonics over the past several billion years.

Since then, Earth has been forming secondary and tertiary crust. Secondary crust forms at mid-ocean spreading centers, where partial-melting of the underlying mantle yields basaltic magmas and new ocean crust forms. This "ridge push" is one of the driving forces of plate tectonics, and it is constantly creating new ocean crust. That means that old crust must be destroyed somewhere, so, opposite a spreading center, there is usually a subduction zone: a trench where an ocean plate is being shoved back into the mantle. This constant process of creating new ocean crust and destroying old ocean crust means that the oldest ocean crust on Earth today is only about 200 million years old.

In contrast, the bulk of the continental crust is much older. The oldest continental crustal rocks on Earth have ages in the range from about 3.7 to 4.28 billion years and have been found in the Narryer Gneiss Terrane in Western Australia, in the Acasta Gneiss in the Northwest Territories on the Canadian Shield, and on other cratonic regions such as those on the Fennoscandian Shield. Some zircon with age as great as 4.3 billion years has been found in the Narryer Gneiss Terrane.

The average age of the current Earth's continental crust has been estimated to be about 2.0 billion years.Most crustal rocks formed before 2.5 billion years ago are located in cratons. Such old continental crust and the underlying mantle asthenosphere are less dense than elsewhere in Earth and so are not readily destroyed by subduction. Formation of new continental crust is linked to periods of intense orogeny; these periods coincide with the formation of the supercontinents such as Rodinia, Pangaea and Gondwana. The crust forms in part by aggregation of island arcs including granite and metamorphic fold belts, and it is preserved in part by depletion of the underlying mantle to form buoyant lithospheric mantle.

Moon's Crust

A theoretical protoplanet named "Theia" is thought to have collided with the forming Earth, and part of the material ejected into space by the collision accreted to form the Moon. As the Moon formed, the outer part of it is thought to have been molten, a "lunar magma ocean." Plagioclase feldspar crystallized in large amounts from this magma ocean and floated toward the surface. The cumulate rocks form much of the crust. The upper part of the crust probably averages about 88% plagioclase (near the lower limit of 90% defined for anorthosite): the lower part of the crust may contain a higher percentage of ferromagnesian minerals such as the pyroxenesand olivine, but even that lower part probably averages about 78% plagioclase. The underlying mantle is denser and olivine-rich.

The thickness of the crust ranges between about 20 and 120 km. Crust on the far side of the Moon averages about 12 km thicker than that on the near side. Estimates of average thickness fall in the range from about 50 to 60 km. Most of this plagioclase-rich crust formed shortly after formation of the moon, between about 4.5 and 4.3 billion years ago. Perhaps 10% or less of the crust consists of igneous rock added after the formation of the initial plagioclase-rich material. The best-characterized and most voluminous of these later additions are the mare basalts formed between about 3.9 and 3.2 billion years ago. Minor volcanism continued after 3.2 billion years, perhaps as recently as 1 billion years ago. There is no evidence of plate tectonics.

Study of the Moon has established that a crust can form on a rocky planetary body significantly smaller than Earth. Although the radius of the Moon is only about a quarter that of Earth, the lunar crust has a significantly greater average thickness. This thick crust formed almost immediately after formation of the Moon. Magmatism continued after the period of intense meteorite impacts ended about 3.9 billion years ago, but igneous rocks younger than 3.9 billion years make up only a minor part of the crust.

Mantle

Mantle Geology

A mantle is a layer inside a planetary body bounded below by a core and above by a crust. Mantles are made of rock or ices, and are generally the largest and most massive layer of the planetary body. Mantles are characteristic of planetary bodies that have undergone differentiation by density. All terrestrial planets (including Earth), a number of asteroids, and some planetary moons have mantles.

Earth's Mantle

The Earth's mantle is a layer of silicate rockbetween the crust and the outer core. Its mass of 4.01

$\times 10^{24}$ kg is 67% the mass of the Earth. It has a thickness of 2,900 kilometres (1,800 mi) making up about 84% of Earth's volume. It is predominantly solid but in geological time it behaves as a viscous fluid. Partial melting of the mantle at mid-ocean ridges produces oceanic crust, and partial melting of the mantle at subduction zones produces continental crust.

Structure

Rheological Structure

The Earth's mantle is divided into two major rheological layers: the rigid lithosphere comprising the uppermost mantle, and the more viscous asthenosphere, separated by the lithosphere-asthenosphere boundary. Lithosphere underlying ocean crust has a thickness of around 100 km, whereas lithosphere underlying continental crust generally has a thickness of 150-200 km. The lithosphere and overlying crust make up tectonic plates, which move over the asthenosphere.

Seismic Structure

The Earth's mantle is divided into three major layers defined by sudden changes in seismic velocity:

- The upper mantle (starting at the Moho, or base of the crust around 7 to 35 km (4.3 to 21.7 mi) downward to 410 km (250 mi)).
- The transition zone (approximately 410–660 km or 250–410 mi), in which wadsleyite (≈ 410–520 km or 250–320 mi) and ringwoodite (≈ 525–660 km or 326–410 mi) are stable.
- The lower mantle (approximately 660–2,891 km or 410–1,796 mi), in which bridgmanite (≈ 660–2,685 km or 410–1,668 mi) and post-perovskite (≈ 2,685–2,891 km or 1,668–1,796 mi) are stable.

The lower ~200 km of the lower mantle constitutes the D" (D-double-prime) layer, a region with anomalous seismic properties. This region also contains LLSVPs and ULVZs.

Mineralogical Structure

The top of the mantle is defined by a sudden increase in seismic velocity, which was first noted by Andrija Mohorovičić in 1909; this boundary is now referred to as the Mohorovičić discontinuity or "Moho".

The upper mantle is dominantly peridotite, composed primarily of variable proportions of the minerals olivine, clinopyroxene, orthopyroxene, and an aluminous phase. The aluminous phase is plagioclase in the uppermost mantle, then spinel, and then garnet below ~100 km. Gradually through the upper mantle, pyroxenes become less stable and transform into majoritic garnet.

At the top of the transition zone, olivine undergoes isochemical phase transitions to wadsleyite and ringwoodite. Unlike nominally anhydrous olivine, these high-pressure olivine polymorphs have a large capacity to store water in their crystal structure. This has led to the hypothesis that the transition zone may host a large quantity of water. At the base of the transition zone, ringwoodite decomposes into bridgmanite (formerly called magnesium silicate perovskite), and ferropericlase. Garnet also becomes unstable at or slightly below the base of the transition zone.

The lower mantle is composed primarily of bridgmanite and ferropericlase, with minor amounts of calcium perovskite, calcium-ferrite structured oxide, and stishovite. In the lowermost ~200 km of the mantle, bridgmanite isochemically transforms into post-perovskite.

Composition

The chemical composition of the mantle is difficult to determine with a high degree of certainty because it is largely inaccessible. Rare exposures of mantle rocks occur in ophiolites, where sections of oceanic lithosphere have been obducted onto a continent. Mantle rocks are also sampled as xenoliths within basalts or kimberlites.

Composition of the Earth's upper mantle (depleted MORB)	
Compound	Mass percent
SiO_2	44.71
Al_2O_3	3.98
FeO	8.18
MnO	0.13
MgO	38.73
CaO	3.17
Na_2O	0.13
Cr_2O_3	0.57
TiO_2	0.13
NiO	0.24
K_2O	0.006
P_2O_5	0.019

Most estimates of the mantle composition are based on rocks that sample only the uppermost mantle. There is debate as to whether the rest of the mantle, especially the lower mantle, has the same bulk composition.The mantle's composition has changed through the Earth's history due to the extraction of magma that solidified to form oceanic crust and continental crust.

Temperature and Pressure

In the mantle, temperatures range from approximately 200 °C (392 °F) at the upper boundary with the crust to approximately 4,000 °C (7,230 °F) at the core-mantle boundary. The geothermal gradient of the mantle increases rapidly in the thermal boundary layers at the top and bottom of the mantle, and increases gradually through the interior of the mantle. Although the higher temperatures far exceed the melting points of the mantle rocks at the surface (about 1200 °C for representative peridotite), the mantle is almost exclusively solid. The enormous lithostatic pressure exerted on the mantle prevents melting, because the temperature at which melting begins (the solidus) increases with pressure.

The pressure in the mantle increases from a few kbar at the Moho to 1390 kbar (139 GPa) at the core-mantle boundary.

Movement

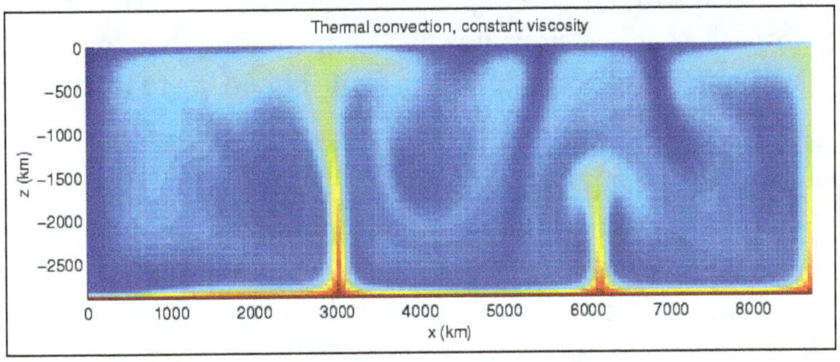

This figure is a snapshot of one time-step in a model of mantle convection. Colors closer to red are hot areas and colors closer to blue are cold areas. In this figure, heat received at the core–mantle boundary results in thermal expansion of the material at the bottom of the model, reducing its density and causing it to send plumes of hot material upwards. Likewise, cooling of material at the surface results in its sinking.

Because of the temperature difference between the Earth's surface and outer core and the ability of the crystalline rocks at high pressure and temperature to undergo slow, creeping, viscous-like deformation over millions of years, there is a convective material circulation in the mantle. Hot material upwells, while cooler (and heavier) material sinks downward. Downward motion of material occurs at convergent plate boundaries called subduction zones. Locations on the surface that lie over plumes are predicted to have high elevation (because of the buoyancy of the hotter, less-dense plume beneath) and to exhibit hot spot volcanism. The volcanism often attributed to deep mantle plumes is alternatively explained by passive extension of the crust, permitting magma to leak to the surface (the "Plate" hypothesis).

The convection of the Earth's mantle is a chaotic process (in the sense of fluid dynamics), which is thought to be an integral part of the motion of plates. Plate motion should not be confused with continental driftwhich applies purely to the movement of the crustal components of the continents. The movements of the lithosphere and the underlying mantle are coupled since descending lithosphere is an essential component of convection in the mantle. The observed continental drift is a complicated relationship between the forces causing oceanic lithosphere to sink and the movements within Earth's mantle.

Although there is a tendency to larger viscosity at greater depth, this relation is far from linear and shows layers with dramatically decreased viscosity, in particular in the upper mantle and at the boundary with the core.The mantle within about 200 km (120 mi) above the core–mantle boundary appears to have distinctly different seismic properties than the mantle at slightly shallower depths; this unusual mantle region just above the core is called D″ ("D double-prime"), a nomenclature introduced over 50 years ago by the geophysicist Keith Bullen. D″ may consist of material from subducted slabs that descended and came to rest at the core–mantle boundary and from a new mineral polymorph discovered in perovskite called post-perovskite.

Earthquakes at shallow depths are a result of strike-slip faulting; however, below about 50 km (31 mi) the hot, high pressure conditions ought to inhibit further seismicity. The mantle is considered to be viscous and incapable of brittle faulting. However, in subduction zones, earthquakes are observed down to 670 km (420 mi). A number of mechanisms have been proposed to explain this phenomenon, including dehydration, thermal runaway, and phase change. The geothermal gradient can be lowered where cool material from the surface sinks downward, increasing the strength of the surrounding mantle, and allowing earthquakes to occur down to a depth of 400 km (250 mi) and 670 km (420 mi).

The pressure at the bottom of the mantle is ~136 GPa (1.4 million atm). Pressure increases as depth increases, since the material beneath has to support the weight of all the material above it. The entire mantle, however, is thought to deform like a fluid on long timescales, with permanent plastic deformation accommodated by the movement of point, line, and planar defects through the solid crystals composing the mantle. Estimates for the viscosity of the upper mantle range between 10^{19} and 10^{24} Pa·s, depending on depth, temperature, composition, state of stress, and numerous other factors. Thus, the upper mantle can only flow very slowly. However, when large forces are applied to the uppermost mantle it can become weaker, and this effect is thought to be important in allowing the formation of tectonic plate boundaries.

Exploration

Exploration of the mantle is generally conducted at the seabed rather than on land because of the relative thinness of the oceanic crust as compared to the significantly thicker continental crust.

The first attempt at mantle exploration, known as Project Mohole, was abandoned in 1966 after repeated failures and cost over-runs. The deepest penetration was approximately 180 m (590 ft). In 2005 an oceanic borehole reached 1,416 metres (4,646 ft) below the sea floor from the ocean drilling vessel *JOIDES Resolution*.

On 5 March 2007, a team of scientists on board the RRS *James Cook* embarked on a voyage to an area of the Atlantic seafloor where the mantle lies exposed without any crust covering, midway between the Cape Verde Islands and the Caribbean Sea. The exposed site lies approximately three kilometres beneath the ocean surface and covers thousands of square kilometres. A relatively difficult attempt to retrieve samples from the Earth's mantle was scheduled for later in 2007. The Chikyu Hakken mission attempted to use the Japanese vessel *Chikyū* to drill up to 7,000 m (23,000 ft) below the seabed. This is nearly three times as deep as preceding oceanic drillings.

A novel method of exploring the uppermost few hundred kilometres of the Earth was proposed in 2005, consisting of a small, dense, heat-generating probe which melts its way down through the crust and mantle while its position and progress are tracked by acoustic signals generated in the rocks. The probe consists of an outer sphere of tungsten about one metre in diameter with a cobalt-60 interior acting as a radioactive heat source. It was calculated that such a probe will reach the oceanic *Moho* in less than 6 months and attain minimum depths of well over 100 km (62 mi) in a few decades beneath both oceanic and continental lithosphere.

Exploration can also be aided through computer simulations of the evolution of the mantle. In 2009, a supercomputer application provided new insight into the distribution of mineral deposits, especially isotopes of iron, from when the mantle developed 4.5 billion years ago.

Other Planetary Mantles

Mercury has a silicate mantle approximately 490 km thick, constituting 28% of its mass. Venus's silicate mantle is approximately 2800 km thick, constituting around 70% of its mass. Mars's silicate mantle is approximately 1600 km thick, constituting ~74-88% of its mass, and may be represented by chassignitemeteorites.

Moons with Mantles

Jupiter's moons Io, Europa, and Ganymede have silicate mantles; Io's ~1100 km silicate mantle is overlain by a volcanic crust, Ganymede's ~1315 km thick silicate mantle is overlain by ~835 km of ice, and Europa's ~1165 km silicate mantle is overlain by ~85 km of ice and possibly liquid water.

The silicate mantle of the Earth's moon is approximately 1300-1400 km thick, and is the source of mare basalts. The lunar mantle might possibly be exposed in the South Pole-Aitken basin and the Crisium basin. The lunar mantle contains a seismic discontinuity at ~500 km depth, most likely related to a change in composition.

Titan and Triton each have a mantle made of ice or other solid volatile substances.

Asteroids with Mantles

Some of the largest asteroids have mantles; for example, Vesta has a silicate mantle similar in composition to diogenite meteorites.

Core

Earth's core is the very hot, very dense center of our planet. The ball-shaped core lies beneath the cool, brittle crust and the mostly-solid mantle. The core is found about 2,900 kilometers (1,802 miles) below Earth's surface, and has a radius of about 3,485 kilometers (2,165 miles).

Planet Earth is older than the core. When Earth was formed about 4.5 billion years ago, it was a uniform ball of hot rock. Radioactive decay and leftover heat from planetary formation (the collision, accretion, and compression of space rocks) caused the ball to get even hotter. Eventually, after about 500 million years, our young planet's temperature heated to the melting point of iron—about 1,538 °Celsius (2,800 °Fahrenheit). This pivotal moment in Earth's history is called the iron catastrophe.

The iron catastrophe allowed greater, more rapid movement of Earth's molten, rocky material. Relatively buoyant material, such as silicates, water, and even air, stayed close to the planet's exterior. These materials became the early mantle and crust. Droplets of iron, nickel, and other heavy metals gravitated to the center of Earth, becoming the early core. This important process is called planetary differentiation.

Earth's core is the furnace of the geothermal gradient. The geothermal gradient measures the increase of heat and pressure in Earth's interior. The geothermal gradient is about 25 °Celsius per kilometer of depth (1 °Fahrenheit per 70 feet). The primary contributors to heat in the core are the decay of radioactive elements, leftover heat from planetary formation, and heat released as the liquid outer core solidifies near its boundary with the inner core.

Unlike the mineral-rich crust and mantle, the core is made almost entirely of metal—specifically, iron and nickel. The shorthand used for the core's iron-nickel alloys is simply the elements' chemical symbols—NiFe.

Elements that dissolve in iron, called siderophiles, are also found in the core. Because these elements are found much more rarely on Earth's crust, many siderophiles are classified as "precious metals." Siderophile elements include gold, platinum, and cobalt.

Another key element in Earth's core is sulfur—in fact 90% of the sulfur on Earth is found in the core. The confirmed discovery of such vast amounts of sulfur helped explain a geologic mystery: If the core was primarily NiFe, why wasn't it heavier? Geoscientists speculated that lighter elements such as oxygen or silicon might have been present. The abundance of sulfur, another relatively light element, explained the conundrum.

Although we know that the core is the hottest part of our planet, its precise temperatures are difficult to determine. The fluctuating temperatures in the core depend on pressure, the rotation of the Earth, and the varying composition of core elements. In general, temperatures range from about 4,400 °Celsius (7,952 °Fahrenheit) to about 6,000 °Celsius (10,800 °Fahrenheit).

Magnetism

Earth's magnetic field is created in the swirling outer core. Magnetism in the outer core is about 50 times stronger than it is on the surface.

It might be easy to think that Earth's magnetism is caused by the big ball of solid iron in the middle. But in the inner core, the temperature is so high the magnetism of iron is altered. Once this temperature, called the Curie point, is reached, the atoms of a substance can no longer align to a magnetic point.

Dynamo Theory

Some geoscientists describe the outer core as Earth's "geodynamo." For a planet to have a geodynamo, it must rotate, it must have a fluid medium in its interior, the fluid must be able to conduct electricity, and it must have an internal energy supply that drives convection in the liquid.

Variations in rotation, conductivity, and heat impact the magnetic field of a geodynamo. Mars, for instance, has a totally solid core and a weak magnetic field. Venus has a liquid core, but rotates too slowly to churn significant convection currents. It, too, has a weak magnetic field. Jupiter, on the other hand, has a liquid core that is constantly swirling due to the planet's rapid rotation.

Earth is the "Goldilocks" geodynamo. It rotates steadily, at a brisk 1,675 kilometers per hour (1,040 miles per hour) at the Equator. Coriolis forces, an artifact of Earth's rotation, cause convection currents to be spiral. The liquid iron in the outer core is an excellent electrical conductor, and creates the electrical currents that drive the magnetic field.

The energy supply that drives convection in the outer core is provided as droplets of liquid iron freeze onto the solid inner core. Solidification releases heat energy. This heat, in turn, makes the remaining liquid iron more buoyant. Warmer liquids spiral upward, while cooler solids spiral downward under intense pressure: convection.

Earth's Magnetic Field

Earth's magnetic field is crucial to life on our planet. It protects the planet from the charged particles of the solar wind. Without the shield of the magnetic field, the solar wind would strip Earth's atmosphere of the ozone layer that protects life from harmful ultraviolet radiation.

Although Earth's magnetic field is generally stable, it fluctuates constantly. As the liquid outer core moves, for instance, it can change the location of the magnetic North and South Poles. The magnetic North Pole moves up to 64 kilometers (40 miles) every year.

Fluctuations in the core can cause Earth's magnetic field to change even more dramatically. Geomagnetic pole reversals, for instance, happen about every 200,000 to 300,000 years. Geomagnetic pole reversals are just what they sound like: a change in the planet's magnetic poles, so that the magnetic North and South Poles are reversed. These "pole flips" are not catastrophic—scientists have noted no real changes in plant or animal life, glacial activity, or volcanic eruptions during previous geomagnetic pole reversals.

The corc is made of two layers: the outer core, which borders the mantle, and the inner core. The boundary separating these regions is called the Bullen discontinuity.

Outer Core

The outer core, about 2,200 kilometers (1,367 miles) thick, is mostly composed of liquid iron and nickel. The NiFe alloy of the outer core is very hot, between 4,500 °C and 5,500 °C (8,132° and 9,932 °Fahrenheit).

The liquid metal of the outer core has very low viscosity, meaning it is easily deformed and malleable. It is the site of violent convection. The churning metal of the outer core creates and sustains Earth's magnetic field.

The hottest part of the core is actually the Bullen discontinuity, where temperatures reach 6,000 °C (10,800 °Fahrenheit)—as hot as the surface of the sun.

Earth's Inner Core

Earth's inner core is the innermost geologic layer of the Earth. It is primarily a solid ball with a radius of about 1,220 kilometres (760 miles), which is about 20% of the Earth's radius and 70% of the Moon's radius.

There are no samples of the Earth's core available for direct measurement, as there are for the Earth's mantle. The information that we have about it mostly comes from analysis of seismic waves and the magnetic field. The inner core is believed to be composed of an iron–nickel alloywith some other elements. The temperature at the inner core's surface is estimated to be approximately 5,700 K (5,430 °C) or 9806 °F, which is about the temperature at the surface of the Sun.

Seismic Waves

Almost all direct measurements that we have about the physical properties of the inner core are the seismic waves that pass through it. The most informative waves are generated by deep earthquakes, 30 km or more below the surface of the Earth (where the mantle is relatively more homogeneous) and recorded by seismographs as they reach the surface, all over the globe.

Seismic waves include "P" (primary or pressure) waves, compressional waves that can travel through solid or liquid materials, and "S" (secondary or shear) shear waves that can only propagate through rigid elastic solids. The two waves have different velocities and are damped at different rates as they travel through the same material.

Of particular interest are the so-called "PKiKP" waves—pressure waves (P) that start near the surface, cross the mantle-core boundary, travel through the core (K), are reflected at the inner core boundary, cross again the liquid core (K), cross back into the mantle, and are detected as pressure waves (P) at the surface. Also of interest are the "PKIKP" waves, that travel through the inner core (I) instead of being reflected at its surface. Those signals are easier to interpret when the path from source to detector is close to a straight line—namely, when the receiver is just above the source for the reflected PKiKP waves, and antipodal to it for the transmitted PKIKP waves.

While S waves cannot reach or leave the inner core as such, P waves can be converted into S waves, and vice-versa, as they hit the boundary between the inner and outer core at an oblique angle. The "PKJKP" waves are similar to the PKIKP waves, but are converted into S-waves when they enter the inner core, travel through it as S-waves (J), and are converted again into P waves when they exit the inner core. Thanks to this phenomeon, it is known that the inner core can propagate S waves, and therefore must be solid.

Other sources of information about the inner core include:

- The magnetic field of the Earth. While it seems to be generated mostly by fluid and electric currents in the outer core, those currents are strongly affected by the presence of the solid inner core and by the heat that flows out of it. (Although made of iron, the core is apparently not ferromagnetic, due to its extremely high temperature).
- The Earth's mass, its gravitational field, and its angular inertia. These are all affected by the density and dimensions of the inner layers.
- The natural oscillation fequencies and modes of the whole Earth oscillations, when large earthquakes make the planet "ring" like a bell. These oscillations too depend strongly on the density, size, and shape of the inner layers.

Physical Properties

Seismic Wave Velocity

The velocity of the S-waves in the core varies smoothly from about 3.7 km/s at the center to about 3.5 km/s at the surface. That is considerably less than the velocity of S-waves in the lower crust (about 4.5 km/s) and less than half the velocity in the deep mantle, just above the outer core (about 7.3 km/s).

The velocity of the P-waves in the core also varies smoothly through the inner core, from about

11.4 km/s at the center to about 11.1 km/s at the surface. Then the speed drops abruptly at the inner-outer core boundary to about 10.4 km/s.

Size and Shape

On the basis of the seismic data, the inner core is estimated to be about 1221 km in radius (2442 km in diameter), which is about 19% of the radius of the Earth and 70% of the radius of the Moon. Its volume is about 7.6 billion cubic km (7.6×10^{18} m^3), which is about 1/140 (0.7%) of the volume of the whole Earth.

Its shape is believed to be very close to an oblate ellipsoid of revolution, like the surface of the earth, only that more spherical: The flattening f is estimated to be between 1/400 and 1/416; meaning that the radius along the Earth's axis is estimated to be about 3 km shorter than the radius at the equator. In comparison, the flattening of the earth as a whole is very close to 1/300, and the polar radius is 21 km shorter than the equatorial one.

Pressure and Gravity

The pressure in the Earth's inner core is slightly higher than it is at the boundary between the outer and inner cores: it ranges from about 330 to 360 gigapascals (3,300,000 to 3,600,000 atm).

The acceleration of gravity at the surface of the inner core can be computed to be 4.3 m/s^2; which is less than half the value at the surface of the Earth (about 10 m/s^2).

Density and Mass

The density of the inner core is believed to vary smoothly from about 13.0 kg/L (= g/cm^3 = t/m^3) at the center to about 12.8 kg/L at the surface. As it happens with other material properties, the density drops suddenly at that surface: the liquid just above the inner core is believed to be significantly less dense, at about 12.1 kg/L. For comparison, the average density in the upper 100 km of the Earth is about 3.4 kg/L.

That density implies a mass of about 10^{23} kg for the inner core, which is 1/60 (1.7%) of the mass of the whole Earth.

Temperature

The temperature of the inner core can be estimated from the melting temperature of impure iron at the pressure which iron is under at the boundary of the inner core (about 330 GPa). From these considerations, in 2002 D. Alfè and other estimated its temperature as between 5,400 K (5,100 °C; 9,300 °F) and 5,700 K (5,400 °C; 9,800 °F). However, in 2013 S. Anzellini and others obtained experimentally a substantially higher temperature for the melting point of iron, 6230 ± 500 K.

Iron can be solid at such high temperatures only because its melting temperature increases dramatically at pressures of that magnitude.

Magnetic Field

In 2010, B. Buffet determined that the average magnetic field in the liquid outer core is about 2.5

milliteslas (25 gauss), which is about 40 times the maximum strength at the surface. He started from the known fact that the Moon and Sun cause tides in the liquid outer core, just as they do on the oceans on the surface. He observed that motion of the liquid through the local magnetic field creates electric currents, that dissipate energy as heat according to Ohm's law. This dissipation in turn dampens the tidal motions and explains previously detected anomalies in Earth's nutation. From the magnitude of the latter effect he could calculate the magnetic field.The field inside the inner core presumably has a similar strength. While indirect, this measurement does not depend significantly on any assumptions about the evolution of the Earth or the composition of the core.

Viscosity

Although seismic waves propagate through the core as if it was solid, the measurements cannot distinguish between a perfectly solid material from an extremely viscous one. Some scientists have therefore considered whether there may be slow convection in the inner core (as is believed to exist in the mantle). That could be an explanation for the anisotropy detected in seismic studies. In 2009, B. Buffett estimated the viscosity of the inner core at 10^{18} Pa·s; which is a sextillion times the viscosity of water, and more than a billion times that of pitch.

Composition

There is still no direct evidence about the composition of the inner core. However, based on the relative prevalence of various chemical elements in the Solar System, the theory of planetary formation, and constraints imposed or implied by the chemistry of the rest of the Earth's volume, the inner core is believed to consist primarily of an iron–nickel alloy.

At the known pressures and estimated temperatures of the core, it is predicted that pure iron could be solid, but its density would exceed the known density of the core by approximately 3%. That result implies the presence of lighter elements in the core, such as silicon, oxygen, or sulfur, in addition to the probable presence of nickel. Recent estimates allow for up to 10% nickel and 2–3% of unidentified lighter elements.

According to computations by D. Alfè and others, the liquid outer core contains 8–13% of oxygen, but as the iron crystallizes out to form the inner core the oxygen is mostly left in the liquid.

Laboratory experiments and analysis of seismic wave velocities seems to indicate that the inner core consists specifically of ε-iron, a crystalline form of the metal with the hexagonal close-packed (hcp) structure. That structure can still admit the inclusion of small amounts of nickel and other elements.

Also, if the inner core grows by precipitation of frozen particles falling onto its surface, then some liquid can also be trapped in the pore spaces. In that case, some of this residual fluid may still persist to some small degree in much of its interior.

Structure

Many scientists had initially expected that the inner core would be found to be homogeneous, because that same process should have proceeded uniformly during its entire formation. It was even suggested that Earth's inner core might be a single crystal of iron.

Axis-aligned Anisotropy

In 1983, G. Poupinet and others observed that the travel time of PKIKP waves (P-waves that travel through the inner core) was about 2 seconds less for straight north-south paths than straight paths on the equatorial plane. Even taking into account the flattening of the Earth at the poles (about 0.33% for the whole Earth, 0.25% for the inner core) and crust and upper mantle heterogeneities, this difference implied that P waves (of a broad range of wavelengths) travel through the inner core about 1% faster in the north-south direction than along directions perpendicular to that.

This P-wave speed anisotropy has been confirmed by later studies, incluing more seismic data and study of the free oscillations of the whole Earth. Some authors have claimed higher values for the difference, up to 4.8%; however, in 2017 D. Frost and B. Romanowicz confirmed that the value is between 0.5% and 1.5%.

Non-axial Anisotropy

Some authors have claim that P-wave speed is faster in directions that are oblique or perpendicular to the N-S axis, at least in some regons of the inner core. However, these claims have been disputed by D. Frost and B. Romanowicz, who instead claim that the direction of maximum speed is as close to the Earth's rotation axis as can be determined.

Causes of Anisotropy

Laboratory data and theoretical computations indicate that the propagation of pressure waves in the hcp crystals of ε-iron are strongly anistropic too, with one "fast" axis and two equally "slow" ones. A preference for the crystals in the core to align in the north-south direction could account for the observed seismic anomaly.

One phenomenon that could cause such partial alignment is slow flow ("creep") inside the inner core, from the equator towards the poles or vice-versa. That flow would cause the crystals to partially reorient themselves accoding to the direction of the flow. In 1996, S. Yoshida and others proposed that such a flow could be caused by higher rate of freezing at the equator than at polar latitudes. An equator-to-pole flow then would set up in the inner core, tending to restore the isostatic equilibrium of its surface.

Others suggested that the required flow could be caused by slow thermal convection inside the inner core. T. Yukutake claimed in 1998 that such convective motions were unliley. However, B. Buffet in 2009 estimated the viscosity of the inner core and found that such convection could have happened, especially when the core was smaller.

On the other hand, M. Bergman in 1997 proposed that the anisotropy was due to an observed tendency of iron crystals to grow faster when their crystallographic axes are aligned with the direction of the cooling heat flow. He therefore proposed that the heat flow out of the inner core would be biased towards the radial direction.

In 1998, S. Karato proposed that changes in the magnetic field might also deform the inner core slowly over time.

Multiple Layers

In 2002, M. Ishii and A. Dziewoński presented evidence that the solid inner core contained an "innermost inner core" (IMIC) with somewhat different properties than the shell around it. The nature of the differences and radius of the IMIC are still unresolved as of 2019, with proposals for the latter ranging from 300 km to 750 km.

A. Wang and X. Song recently proposed a three-layer model, with an "inner inner core" (IIC) with about 500 km radius, an "outer inner core" (OIC) layer about 600 km thick, and an isotropic shell 100 km thick. In this model, the "faster P-wave" direction would be parallel to the Earth's axis in the OIC, but perpendicular to that axis in the IIC. However, conclusion has been disputed by claims that there need not be sharp discontinuities in the inner core, only a gradual change of properties with depth.

Lateral Variation

In 1997, S. Tanaka and H. Hamaguchi claimed, on the basis of seismic data, that the anisotropy of the inner core material, while oriented N-S, was more pronounced in "eastern" hemisphere of the inner core (at about 110 °E longitude, roughly under Borneo) than in the "western" hemisphere (at about 70 °W, roughly under Colombia.

Alboussère and others proposed that this asymmetry could be due to melting in the Eastern hemisphere and re-crystallization in the Western one. C. Finlay conjectured that this process could explain the asymmetry in the Earth's magnetic field.

However, in 2017 D. Frost and B. Romanowicz disputed those earlier inferences, claiming that the data shows only a weak anisotropy, with the speed in the N-S direction being only 0.5 to 1.5% faster than in equatorial directions, and no clear signs of E-W variation.

Other Structure

Other researchers claim that the properties of the inner core's surface vary from place to place across distances as small as 1 km. This variation is surprising, since lateral temperature variations along the inner-core boundary are known to be extremely small (this conclusion is confidently constrained by magnetic field observations).

Growth

The Earth's inner core is thought to be slowly growing as the liquid outer core at the boundary with the inner core cools and solidifies due to the gradual cooling of the Earth's interior (about 100 degrees Celsius per billion years).

According to calculations by Alfé and others, as the iron crystallizes onto the inner core, the liquid just above it becomes enriched in oxygen, and therefore less dense than the rest of the outer core. This process creates convection currents in the outer core, which are thought to be the prime driver for the currents that create the Earth's magnetic field.

The existence of the inner core also affects the dynamic motions of liquid in the outer core, and thus may help fix the magnetic field.

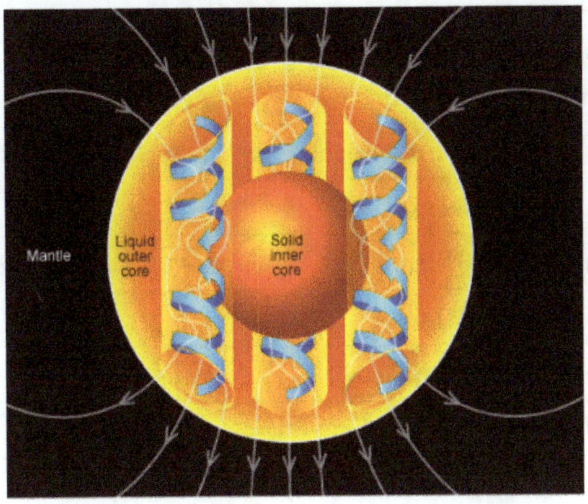

Schematic of the Earth's inner core and outer core motion and the magnetic field it generates.

Dynamics

Because the inner core is not rigidly connected to the Earth's solid mantle, the possibility that it rotates slightly more quickly or slowly than the rest of Earth has long been entertained. In the 1990s, seismologists made various claims about detecting this kind of super-rotation by observing changes in the characteristics of seismic waves passing through the inner core over several decades, using the aforementioned property that it transmits waves more quickly in some directions. In 1996, X. Song and P. Richards estimated this "super-rotation" of the inner core relative to the mantle as about one degree per year. In 2005, they and J. Zhang compared recordings of "seismic doublets" (recordings by the same station of earthquakes occurring in the same location on the opposite side of the Earth, years apart), and revised that estimate to 0.3 to 0.5 degree per year.

In 1999, M. Greff-Lefftz and H. Legros noted that the gravitational fields of the Sun and Moon that are responsible for ocean tides also apply torques to the Earth, affecting its axis of rotation and a slowing down of its rotation rate. Those torques are felt mainly by the crust and mantle, so that their rotation axis and speed may differ from overall rotation of the fluid in the outer core and the rotation of the inner core. The dynamics is complicated because of the currents and magnetic fields in the inner core. They find that the axis of the inner core wobbles (nutates) slightly with a period of about 1 day. With some assumptions on the evolution of the Earth, they conclude that the fluid motions in the outer core would have entered resonance with the tidal forces at several times in the past (3.0, 1.8, and 0.3 billion years ago). During those epochs, which lasted 200–300 million years each, the extra heat generated by stronger fluid motions might have stopped the growth of the inner core.

Age

Theories about the age of the core are necessarily part of theories of the history of Earth as a whole. This has been a long debated topic and is still under discussion at the present time. It is widely believed that the Earth's solid inner core formed out of an initially completely liquid core as the Earth cooled down. However, there is still no firm evidence about the time when this process started.

Age estimates from different studies and methods			
Date	Authors	Age	Method
2001	Labrosse et al.	1±0.5	T(N)
2003	Labrosse	~2	T(R)
2011	Smirnov et al.	2–3.5	P
2015	Labrosse	< 0.7	T
2015	Biggin et al.	1–1.5	P
2016	Ohta et al.	< 0.7	T
2016	Konôpková et al.	< 4.2	T
2019	Bono et al.	0.5	P
T = thermodynamic modeling P = paleomagnetism analysis (R) = with radioactive elements (N) = without them			

Two main approaches have been used to infer the age of the inner core: thermodynamic modeling of the cooling of the Earth, and analysis of paleomagnetic evidence. The estimates yielded by these methods still vary over a large range, from 0.5 to 2 billion years old.

Thermodynamic Evidence

One of the ways to estimate the age of the inner core is by modeling the cooling of the Earth, constrained by a minimum value for the heat flux at the core–mantle boundary (CMB). That estimate is based on the prevailing theory that the Earth's magnetic field is primarily triggered by convectioncurrents in the liquid part of the core, and the fact that a minimum heat flux is required to sustain those currents. The heat flux at the CMB at present time can be reliably estimated because it is related to the measured heat flux at Earth's surface and to the measured rate of mantle convection.

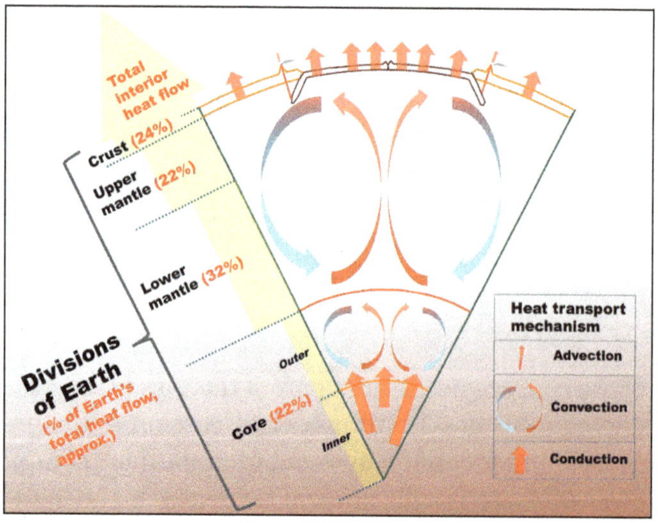

Heat flow of the inner earth.

In 2001, S. Labrosse and others, assuming that there were no radioactive elements in the core, gave an estimate of 1±0.5 billion years for the age of the inner core — considerably less than the estimated age of the Earth and of its liquid core (about 4.5 billion years) In 2003, the same group concluded that, if the core contained a reasonable amount of radioactive elements, the inner core's age could be a few hundred million years older.

In 2012, theoretical computations by M. Pozzo and others indicated that the electrical conductivity of iron and other hypothetical core materials, at the high pressures and temperatures expected there, were two or three times higher than assumed in previous research. These predictions were confirmed in 2013 by measurements by Gomi and others. The higher values for electrical conductivity led to increased estimates of the thermal conductivity, to 90 W/m/K; which, in turn, lowered estimates of its age to less than 700 million years old.

However, in 2016 Konôpková and others directly measured the thermal conductivity of solid iron at inner core conditions, and obtained a much lower value, 18–44 W/m/K. With those values, they obtained an upper bound of 4.2 billion years for the age of the inner core, compatible with the paleomagnetic evidence.

In 2014, Driscoll and Bercovici published a thermal history of the Earth that avoided the so-called mantle thermal catastrophe and new core paradox by invoking 3 TW of radiogenic heating by the decay of 40K in the core. Such high abundances of K in the core are not supported by experimental partitioning studies, so such a thermal history remains highly debatable.

Paleomagnetic Evidence

Another way to estimate the age of the Earth is to analyze changes in the magnetic field of Earth during its history, as trapped in rocks that formed at various times (the "paleomagnetic record"). The presence or absence of the solid inner core could result in very different dynamic processes in the core which could lead to noticeable changes in the magnetic field.

In 2011, Smirnov and others published an analysis of the paleomagnetism in a large sample of rocks that formed in the Neoarchean (2.8 to 2.5 billion years ago) and the Proterozoic (2.5 to 0.541 billion). They found that the geomagnetic field was closer to that of a magnetic dipole during the Neoarchean than after it. They interpreted that change as evidence that the dynamo effect was more deeply seated in the core during that epoch, whereas in the later time currents closer to the core-mantle boundary grew in importance. They further speculate that the change my have been due to growth of the solid inner core between 3.5 and 2 billion years ago.

In 2015, Biggin and others published the analysis of an extensive and carefully selected set of Pre-cambriansamples and observed a prominent increase in the Earth's magnetic field strength and variance around 1–1.5 billion years ago. This change had not been noticed before due to the lack of sufficient robust measurements. They speculated that the change could be due to the birth of Earth's solid inner core. From their age estimate they derived a rather modest value for the thermal conductivity of the outer core, that allowed for simpler models of the Earth's thermal evolution.

An analysis of rock samples from the Ediacaran epoch (formed about 565 million years ago), published by Bono and others in 2019, revealed unusually low intensity and two distinct directions for the geomagnetic field during that time. Considering other evidence of high frequency of magnetic field reversals around that time, they speculate that those anomalies could be due to the onset of formation of the inner core, which would then be 0.5 billion years old.

Faults

Faults are fractures in the earth's crust along which slipping has occurred parallel to the surface of the fracture. Some are clean sharp breaks. Many however, consist of sub parallel faults among which the total displacements have been distributed.

The terms shear zone ox fault zone are often used to closely spaced sub-parallel structures along which there has been distributive movement. Some faults are knife like breaks. Other faults, due to the frictional effects of rock masses sliding over one another break or crack the rock on either side of the rupture.

Some faults pulverize the rock in the fault zone to clay-like powder called gouge. By convention the surface of rupture along which relative movements have occurred are termed fault planes. The fault surfaces in most cases are warped or curved and irregular and hence the term fault surface is preferable to fault plane.

Classification and Types of Faults

The major types of faults are the following:

1. Normal Fault: A normal fault is a dip slip fault in which the hanging wall slips downward relative to the footwall. This fault occurs along a steep fault plane with a hade of 10° to 20°. The throw i.e. the vertical component of the movement is large. This fault occurs due to stretching of rocks.

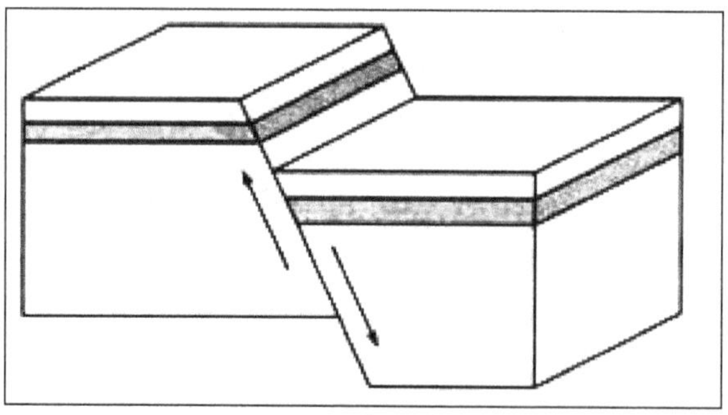

Normal Fault.

2. Reverse Fault: A reverse fault is the result of severe compressive stresses, where the hanging wall moves up the fault plane relative to the footwall. The two sides of the fault move closer together. This fault is also called a thrust fault. The fault plane is low angled resulting in a large horizontal movement.

3. Strike-Slip Fault: In this type of fault, the separated blocks on either sides of the fault plane move shearing off in the direction of the strike. This fault is also called lateral fault, trans-current fault, wrench fault or tear fault.

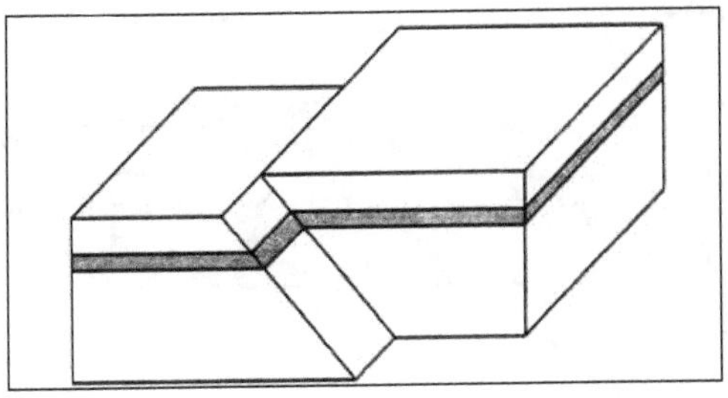

Strike-Slip Fault.

4. Step Fault: This is a fault system consisting of a number of faults with parallel fault planes, the separated blocks slipping in the same direction along parallel planes giving a step-like feature. This fault is also called a fault terrace.

Faulted Blocks Showing the Cahracteristics Displacement for the Different Classes of Faults.

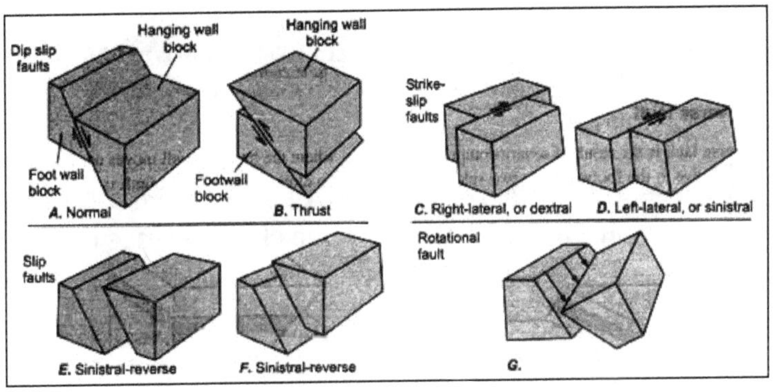

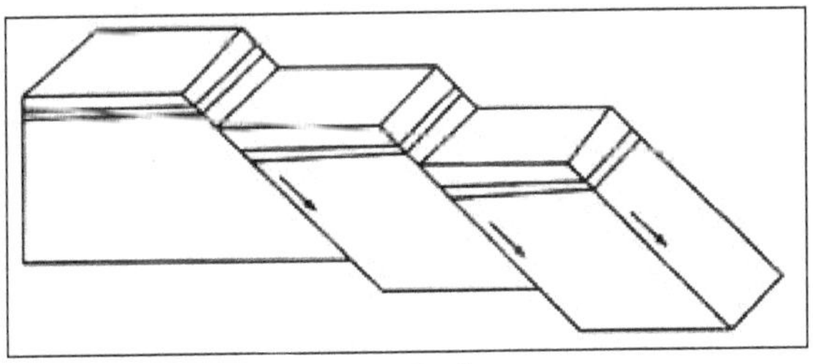

Step Fault.

5. Trough Fault: This is a fault system in which two normal faults occur whose fault planes are inclined providing a common down throw side between them. The downthrown block between the two fault planes forms a long trench called a graben or a rift valley. Some grabens are very long and also deep and may be filled with water to become a water stream.

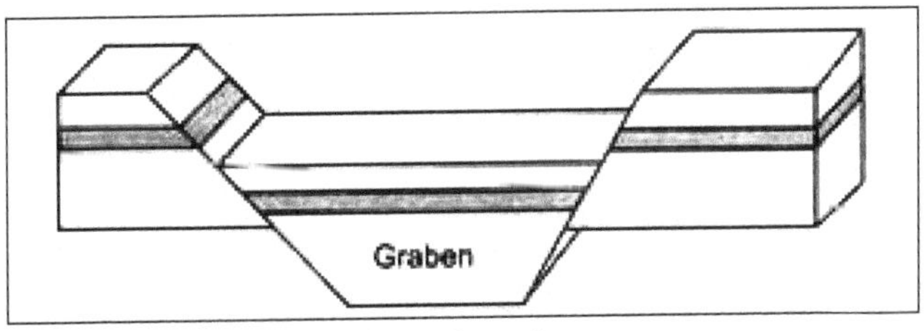

Trough Fault.

6. Ridge Fault: In this case two normal faults occur whose fault planes are providing a common upthrow between them.

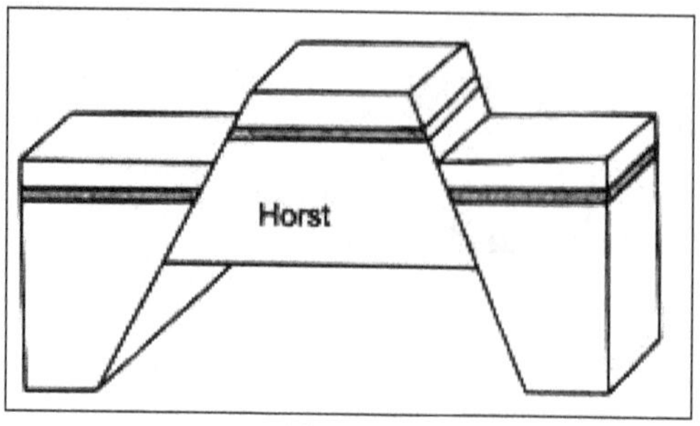

Ridge Fault.

The up thrown middle wedge shaped block forms a long ridge between the fault planes and is called a horst or & fault ridge.

Fault Scarps

Many faults are known to break through the ground surface as well as the rocks beneath. When the part on one side of the fault plane moves up with respect to the other it will result in the form of a cliff or fault scarp.

The height of such a cliff observed at the present time depends on not only the extent of displacement during the fault but also on the long period of time elapsed since the fault occurred. At many places erosion can reduce the projecting part of the up throw block to the same level as that of the surface of the down throw block. As a result different rock types or rock strata may be exposed at the surface on either side of the fault.

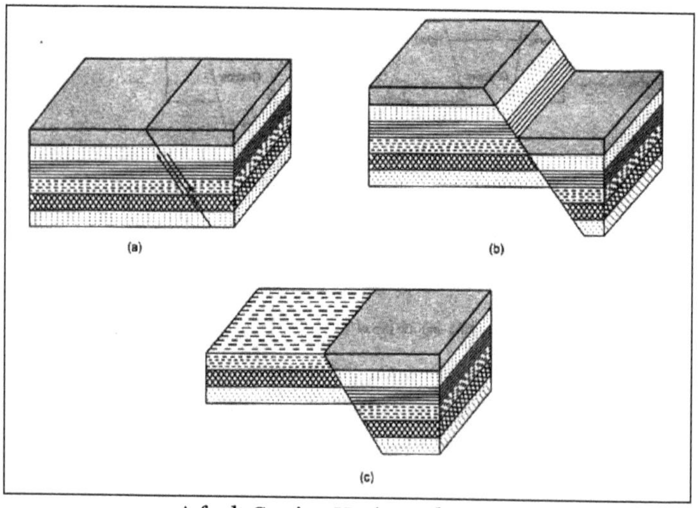

A fault Cutting Horizontal Strata.

Rate of Faulting

At some places displacements of a few centimetres to 6 metres have occurred in a few minutes. Such sudden movements mostly produce earthquakes. There also exist places where faulting takes place continually at a very slow rate. Though over a long period of time the total displacement may be great, the surface changes may escape detection, since the up-thrown part is continuously eroded as fast as it is elevated.

Horsts and Grabens

In some situations a few normal faults occur whose fault planes mutually diverge or converge displacing wedge shaped blocks upwards or displacing blocks downwards as shown in. The alternating raised blocks are called Horsts and the alternating lowered blocks are called Grabens.

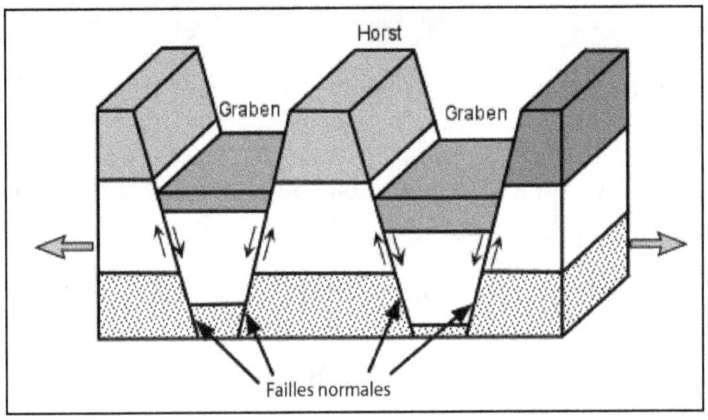

Horsts and Grabens.

Horsts form long ridges and grabens form long troughs. Grabens may get filled with water forming ponds, lakes and streams.

Distinguishing Faults from Geometric Relations

The following geometric relations of the fault to the strata should be noted to easily distinguish the type of fault:

1. In a bedding fault the strike and the dip of the fault are both parallel to the strike and dip of the strata.
2. In a strike fault, the strike of the fault is parallel to the strike of the strata, but the dip of the fault cuts across the dip of the strata.
3. In a dip fault the strike of the fault is perpendicular to the strike of the strata.
4. In an oblique fault, the strike of the fault is oblique forming a distinct acute angle with the strike of the strata.

In a bedding fault, the plane of movement is the bedding plane of the strata. Hence the displacement can be in any direction in that plane. It is not shown by strata which maintain their continuity and parallelism but can be detected by striations by offset joints or dikes.

Strike faults exhibit a repetition of strata or a gap in cross section according to the manner in which the dip of the fault cuts across the dip of the strata and according to the normal or reverse type of the fault.

Dip faults cut off the strata across their strike and offset them to one side or the other. The obvious displacement is a horizontal one, but it may be occasioned by a vertical movement across inclined beds.

Oblique faults approximate the effects of strike faults or of dip faults, depending on whether they strike more nearly parallel with the strike of the strata or perpendicular to it.

Any of these four types of faults (bedding, strike, dip or oblique faults, may be either normal or reverse faults. They may have a displacement parallel to the strike of the fault or perpendicular to it.

The direction of displacement, whether parallel, perpendicular or oblique to the strike of the fault is a further means of classification of faults according to their geometric relations.

On this basis:

1. In a strike slip fault one wall has slipped past the other in a direction parallel to the strike of the fault.
2. In a dip-slip fault, one wall has slipped past the other down the dip of the fault or has been pushed past the other up the dip of the fault.
3. In an oblique-slip fault the walls are displaced in a direvction oblique to the strike of the fault.

Effects of Faulting on Geologic or Stratigraphic Units

Generally displacements along faults, places adjacent to one another, rocks which do not belong together in ordinary geologic sequences. The discontinuity that results gives an indication of the presence of a fault.

In a continuous geologic feature like a sedimentary bedding if a break in the feature is present it is an indication of the presence of a fault.

Presence of horses or fault slices along a discontinuity is a clear evidence of a fault. Horses consist of volumes of rock bounded on all sides by faults. They are sliced from either the footwall or the hanging wall block by a branch of the fault and are displaced at a considerable distance from their original position.

They may appear remarkably out of place stratigraphically. In areas where horses separate two similar rock types, a horse of a different lithology may be the only evidence noticeable for a fault.

Omission or repetition of strata in a known stratigraphic sequence is yet another indication of a fault. Faults have pronounced effects on topography, stream channels and groundwater flow. These effects suggest the presence of a fault.

Scarps are linear features distinguished by sharp increases in the topographic slope and suggest the presence of a fault.

Fault benches are linear topographic features distinguished by noticeable decrease in slope. This feature occurs where a fault displaces an originally existing smooth slope to form a strip of shallower slope. It is also possible that erosion of less resistant rocks in a fault zone can also produce a shallower slope than the surroundings with more resistant rocks.

Ridges, valleys and streams may be offset along a fault. The deflection of a stream channel gives an indication of slip on the fault.

A fault surface or a fault zone can act either as a conduit or a barrier for groundwater depending on the permeability of the material in the fault and on either side of the fault. If breccia is present it serves as an excellent conduit for water but if thick gouge zone with clay minerals is present, it serves as a barrier to flow of water. Faults may offset an aquifer and disturb the ground water flow.

- Evidence of Faulting: Often many faults are difficult to be detected in the field. Many faults shown in geological maps are based on result of inferences rather than direct observations. These observations may be divided into two groups viz., lithologic which suggest or establish faults and physiographic evidence.
- Lithologic Evidence: There are many varieties of lithologic features related to faulting. The significant among these are slicken sides, brecciation and gouge, shear zones, displacements and drag.
- Slicken Sides: These are parallel striations or grooves displayed by the fracture surface on which movements have taken place. Such a slicken sided surface is usually well polished due to frictional rubbing of one block by the other.

  The direction of movement is indicated by the trend of the striae, and the direction of relative displacements can be determined from many slicken sided and polished surfaces by passing the hand over the surface to find the rough and smooth directions.
- Brecciation: The rocks are highly fractured or even crushed to angular fragments along the faults and are called breccias. An elongated zone of brecciation which intersects bedding is suggestive of faulting. The fragments constituting a fault breccia are extremely variable. The fragments may be as large as 2 or 3 metres and may be of minute size too. A very fine clay-like product of fault crushing is called gouge.
- Shear zones: In many cases, faults are characterized by closely spaced fractures among which movements have been distributed. Shear zones are suggestive evidence of faults. At many places, due to weather action along the fracture zone is more advanced than in the adjacent rock. Much of the difficulty in engineering construction in fault areas arises from the altered or rotten rock encountered.

  Since fractures provide avenues of ready percolation for circulating waters, many mineral deposits are localized along faults. Some faults have shear zones and are silicified by more or less complete replacement along the zone or by a network of quartz veins which fill the fractures.
- Drag: Drag refers to the minor folding of strata along the walls of a fault produced by the fault displacement. For instance in an area of regular structural attitude, one of the horizontal beds may abruptly change in attitude indicating the drag associated with faulting. In zones of completely folded rock evidence of drags loses importance.
- Dislocations: It is possible to observe actual dislocation of strata, veins or dikes and to match the ends of dislocated parts along some faults of small displacements. The repetition or missing of recognizable beds often establishes the break.

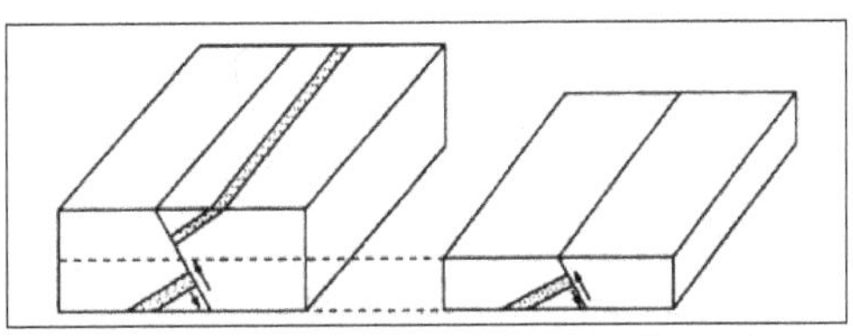

Elimination of Bed by Faulting and Erosion.

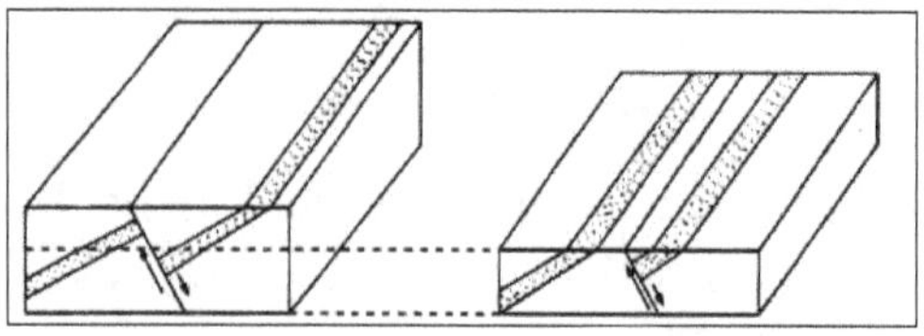

Repitition of Beds by Faulting and Erosion.

An abrupt termination of structures like folds, beds or dikes along a common line or zone indicates faults.

- Physiographic Evidences: Physiographic features or landscape forms may also be indicative of faulting when seen in the field or recognized on maps or aerial photographs. Escarpments and other suggestive topographic features are of interest in locating faults.

  Escarpments are linear forms sudden increase of slope. Faulting can lead to two types of escarpments, viz., fault scarps and fault-line scarps.

  An actual surface of displacement may stand up as an escarpment unmodified by erosion. This form of escarpment constitutes a fault scarp. The inclination of the slope varies more than the dip of the fault since erosion softens or flattens the slope. Triangular facets as shown in are locally found due to dissection of fault scarps.

  Fault scarps are found only where faulting has been geologically very recent. Earthquake centres at escarpments are indicating evidence of fault origin. Escarpments in unconsolidated deposits like alluvial fans or lake fills are also indicative of recent faults.

  Fault-line scarps are those etched out along the break by subsequent erosion. Frequently faults bring together resistant as well as non-resistant rocks. Continuous erosion in a long interval of time leaves the hard and more resistant rocks in relief, thus creating a linear scarp along a fault zone.

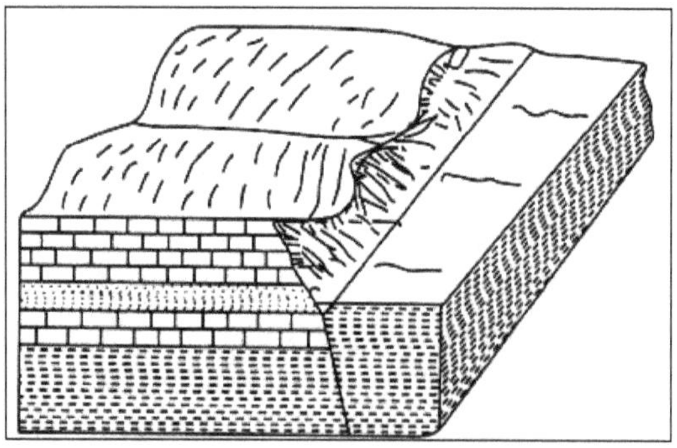

Triangula Facets Caused by erosion of Fault Scrap.

Importance of Faults

Faults are important due to their various dangerous as well as useful effects.

These are briefly given below:

- Faults cause considerable damage to rocks and are therefore real hazards in mining and engineering works.

- Fault breccia and fault gouge (pulverized rocks) have low strength and are poor foundation materials.
- There is additional difficulty and additional expenditure in excavations since fractured rocks are difficult to handle.
- Earthquakes and landslides are likely to be triggered by faults.
- To provide earthquake proof structures it is necessary to have a knowledge of the location and earthquake potential of faults. This is of particular importance in locating sites for dams and reservoirs, hydropower structures, underground power, nuclear power stations, tunnels, public buildings, school and college buildings etc.
- Filling big reservoirs behind dams across river valleys may trigger movements along fault planes leading to earthquakes.
- Faults may create passages for water percolation. They may also provide passages for mineralising. Many fault zones are sites of mineralization.
- Fault throw and fault heave constitute most salient factors in the exploration and recovery of mineral veins and coal seams.
- Faults may create lakes, swamps and marshy zones.
- Some fault zones form potential oil traps.

Folds

Ductile deformation of a layered rock forms bends or warps called folds. Folding is caused due to compressive stresses. When a layered rock folds, it crinkles similar to bunched up fabric. Folds generally occur at great depths in the earth's crust where the rock layers are exposed to high temperatures and pressures.

When these rocks are compressed by convergent plate movements they respond plastically by crumpling or folding. In many cases folds occur in series. They generally cover great lengths as well as breadths. The radius of curvature of the fold is in most cases small compared with the amplitude and wave length of the wave shaped fold. A fold in some cases may also be a single local bend.

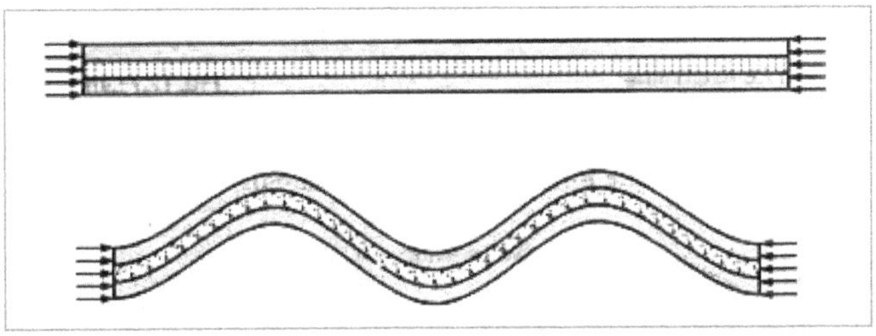

Long Straight Rock Strata Developing Folds under the Action of Compressive Forces.

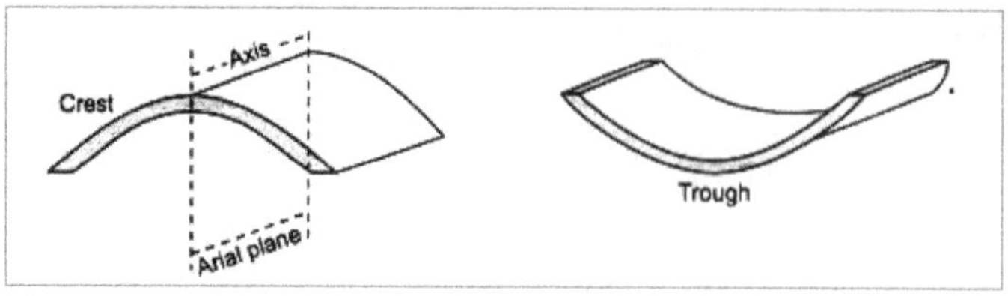

Axis.

Parts of a Fold

The various parts of a fold are briefly described below:

- Axis: This is the median line about which the strata has folded. The axis of a fold may be horizontal, inclined or vertical. If the axis of the fold is horizontal then it is parallel to the strike of the fold.
- Axial Plane: This is a plane dividing the fold into two somewhat symmetrical parts. The axial plane may be vertical, inclined or horizontal. The line of intersection of the axial plane with any bedding plane is called the fold axis or the axial line.

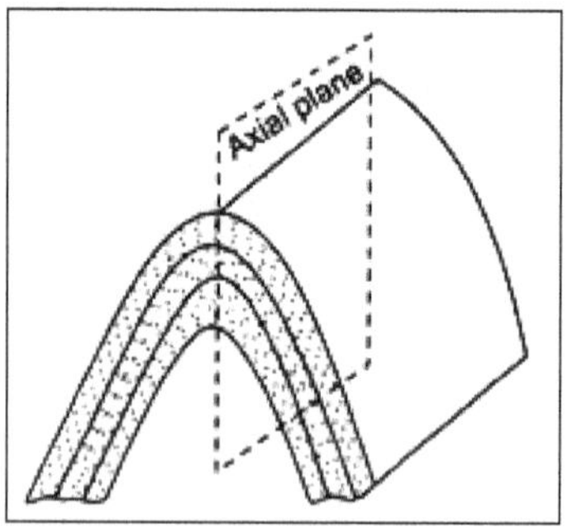

Axial Plane.

- Limbs of a Fold: Two sides of a fold (on either side of the axial plane) are called the limbs of the fold. In the usual cases the limbs are inclined. However in particular cases they may be vertical or horizontal. The limbs may dip either way from each other or towards each other. The angle between the limbs is called the inter-limb angle.

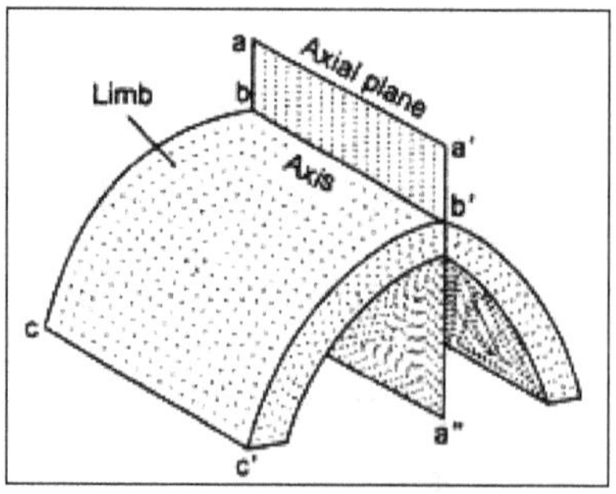

Limbs of a Fold.

- Crest of a Fold: This the upper portion of an up-fold.
- Trough of a Fold: This is the lower portion of a down-fold.
- Wavelength of a Fold: This is the distance between successive crests or troughs of the fold.

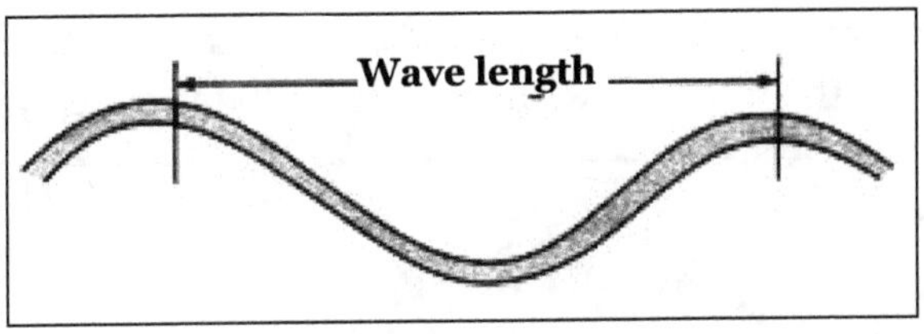

Wavelength of a Fold.

Classification of Folds

Folds are classified into two main types namely anticlines or up-folds and synclines or down-folds.

- Anticline Folds: An anticline consists of beds bent upwards with limbs dipping away from each other.

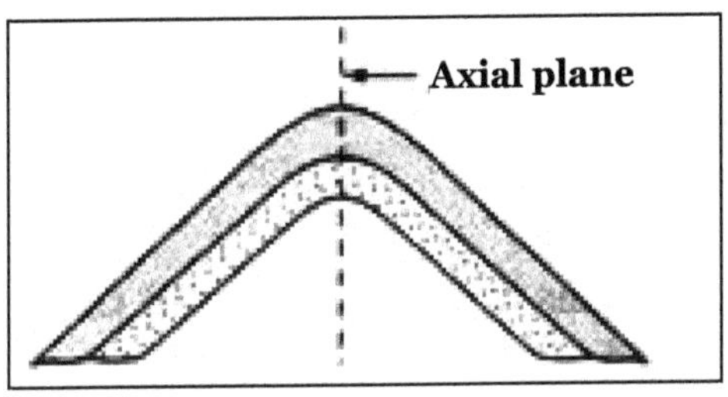

Anticline.

- Syncline Folds: A syncline consists of beds bent downwards with limbs dipping towards each other.

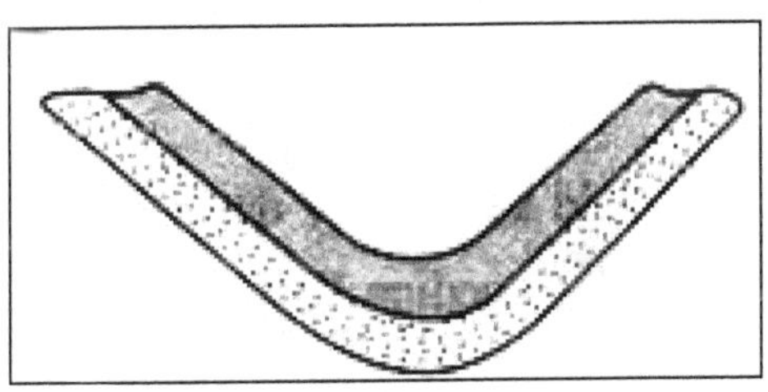

Syncline.

- Symmetrical Fold and Asymmetrical Fold: A symmetrical fold is a fold whose axial plane is vertical and the limbs dip equally. The axial plane in this case divides the fold into two equal halves. If the two limbs dip at different angles the fold is an Asymmetrical fold.
- Monocline: This is a fold in which only one limb is bent. This is a case when a rock-bed bends abruptly and resumes the original attitude at the lower level.

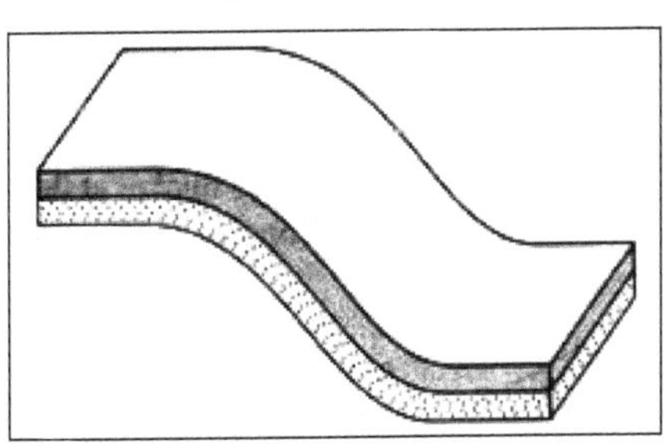

Monocline.

- Plunging Fold or Pitching Fold: This is a fold whose axis is at some angle with the horizontal. The inclination of the fold axis with the horizontal is called plunge of the fold.

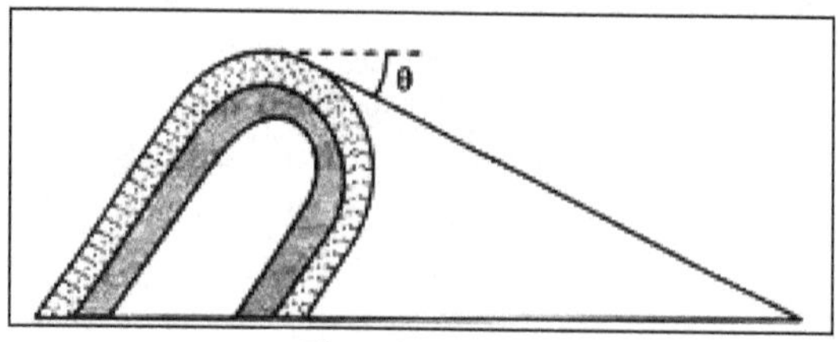

Plunging Fold.

- Isocline or Carinate Fold: This is a fold whose limbs dip at the same angle in the same direction. The two limbs in this case are parallel. The axial plane may be vertical, inclined or horizontal.

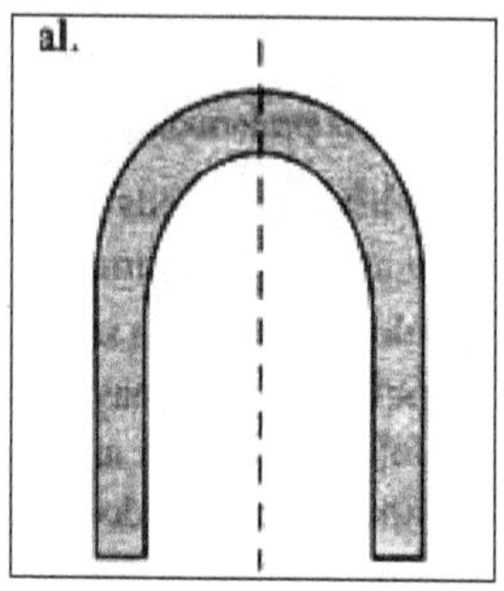

Vertical Isocline.

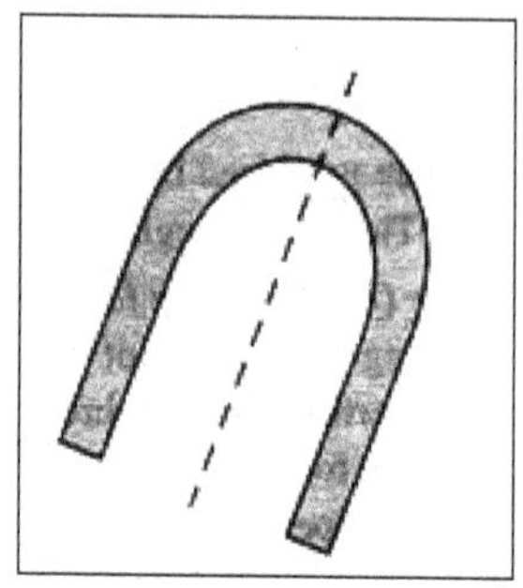

Inclined Isocline.

- Overturned Fold: This is a fold whose limbs dip unequally in the same direction.

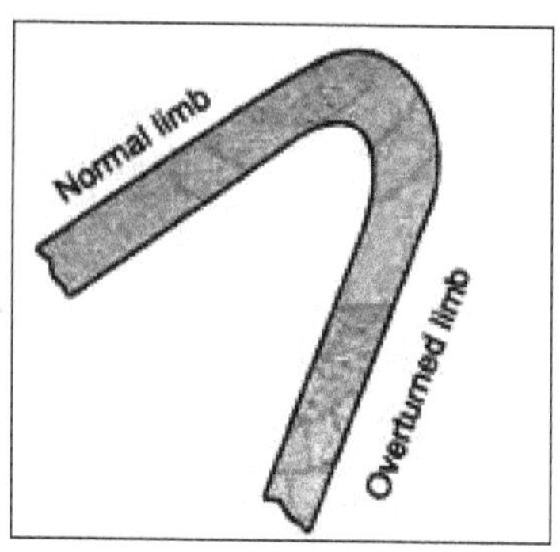

Overturned Fold.

- Recumbent Fold: This is a fold whose limbs are bent back on themselves almost horizontally.

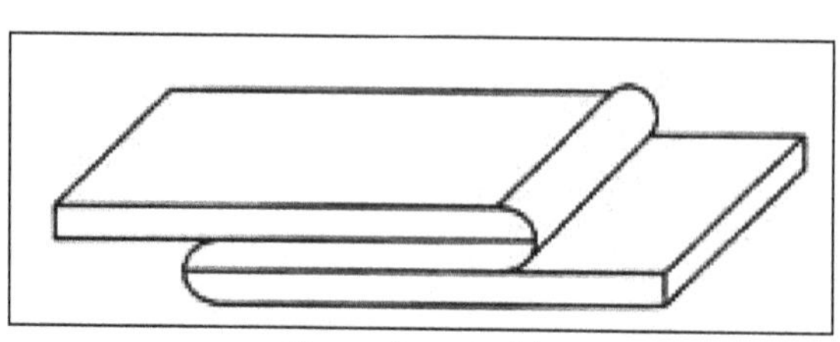

Recumbent Fold.

- Zigzag Fold or Chevron Fold: This is a fold having a sharp angular crest or trough.

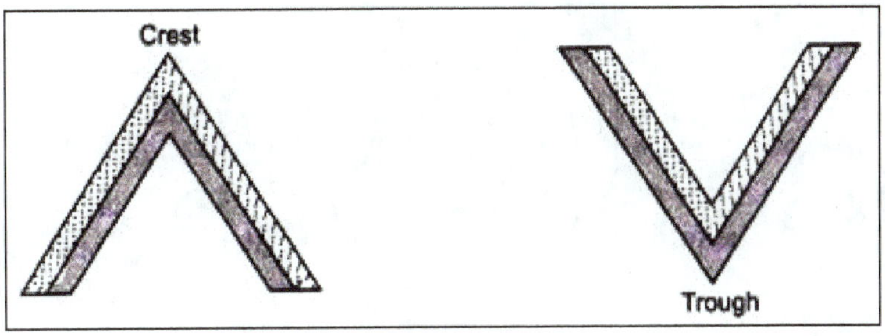

Chevron Fold.

- Supratenuous Fold: This is a fold whose beds are thinner at the crest and thicker at the trough. Such folds are formed due to contemporaneous sedimentation, compaction over irregular surfaces uplift folding, sinking etc.

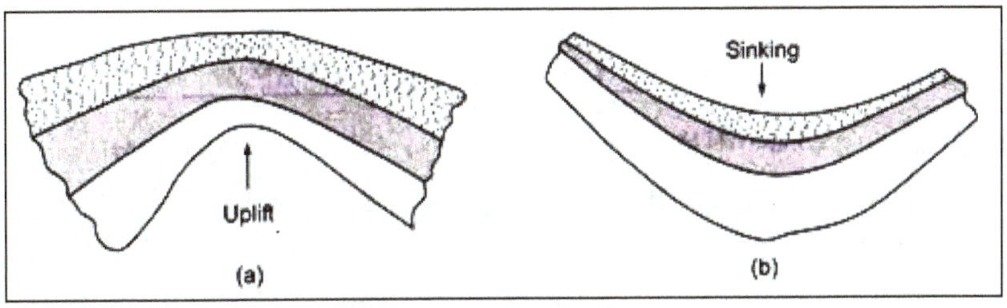

Supratenuous Fold.

These folds are produced by tangential pressures which lift up the beds slowly and vertically at the crests. The thick troughs are formed due to sinking and large accumulation of sediments.

- Dome Fold or Quaquaversal Fold or Pericline: Dome fold consists of a set of rock beds lifted centrally giving the feature of a dome. The area of rock bed lifted may be circular or oval shaped. In a vertical section through the summit, the fold exhibits an anticlinal feature. For this reason this fold is also called a compound anticline. After the domes are eroded, the younger rocks appear surrounding the older rocks.

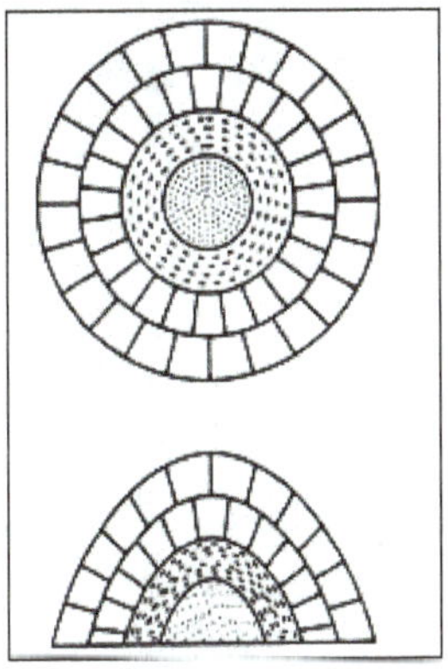

Dome Fold.

- Basin Fold or Centrocline: Basin fold consists of a set of rock beds which are sunk down centrally giving the feature of a basin. The area of the rock bed sunk may be circular or oval shaped. In a vertical section taken centrally the fold exhibits a synclinal feature. For this reason this fold is also called a compound syncline.

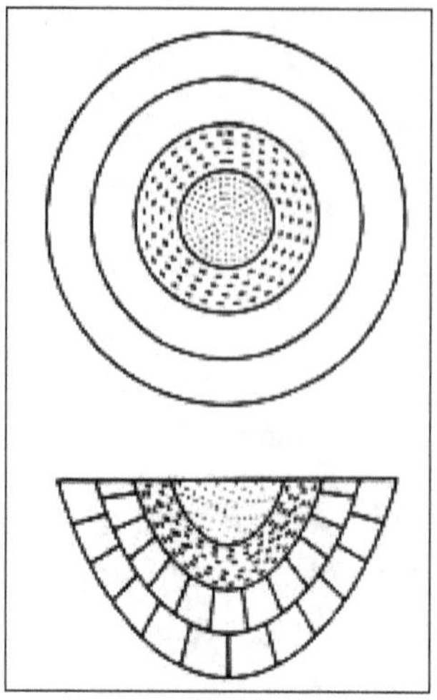

Basin Fold.

Fold Systems

A rock strata is unlikely to form a single fold. In most cases it forms a series or a group of folds. Such a series of folds brought about by the same earth movement extending to great distances is called a fold system. A fold system consisting of a number of symmetrical anticlines and synclines, with their axes horizontal and parallel is called a non-plunging fold system.

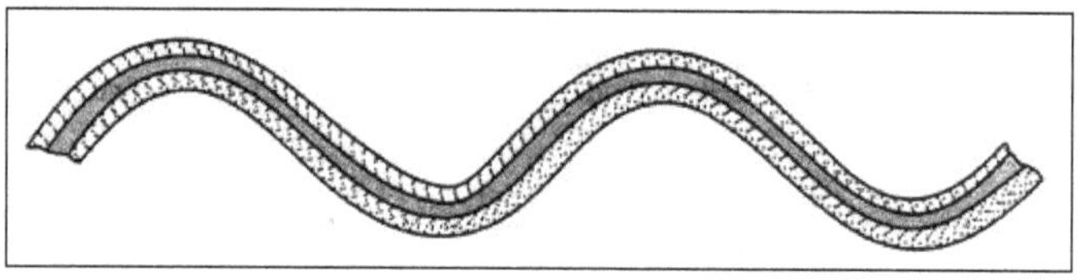

Non-Plunging Fold System.

- Plunging Fold System: This is a fold system with plunging anticlines and synclines. The limbs of the folds dip in one direction.

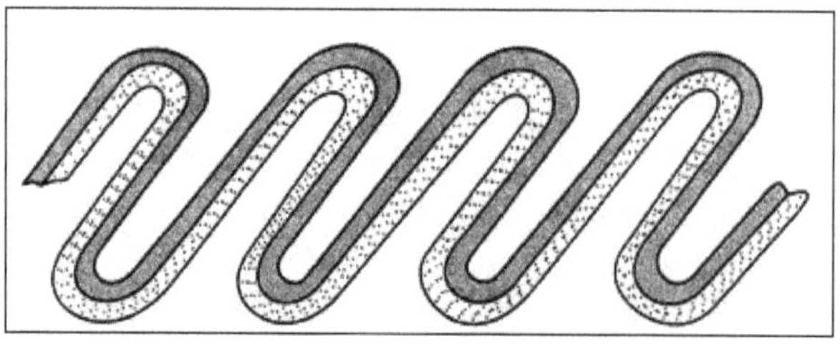

Plunging Fold System.

- Anticlinorium: This is a system of anticlines and synclines which are all arched up into a major up fold.

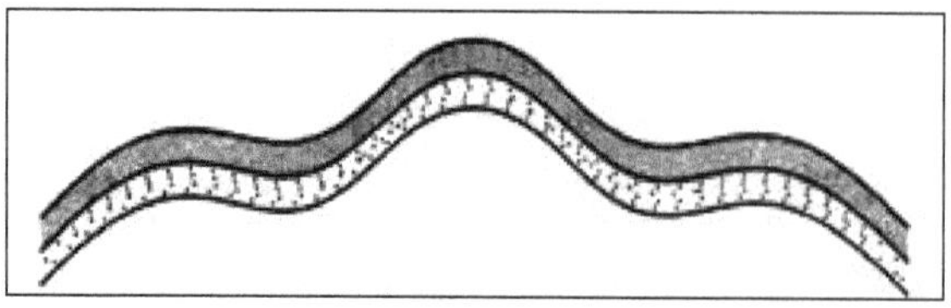

Anticlinorium.

- Synclinorium: This is a system of anticlines and synclines which are all arched down into a major down fold.

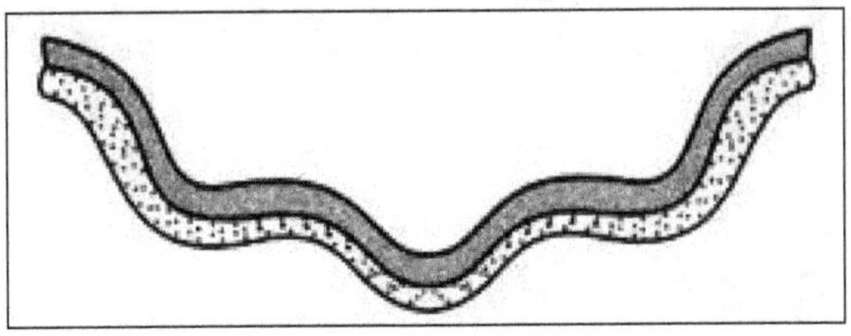

Synclinorium.

- Fan Fold: This is a system of folds consisting of anticlines and synclines but with limbs having contrasting features. In this case the limbs of anticlines dip towards each other and the limbs of synclines dip away from each other.

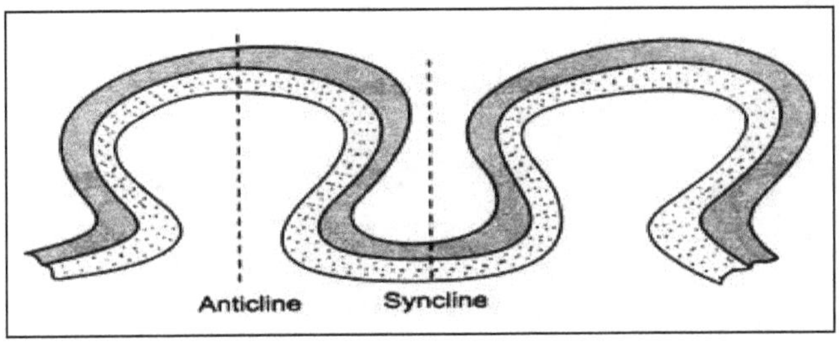

References

- Advance-geology-notes-on-geology- 91078: yourarticlelibrary.com, Retrieved 05 January, 2019
- 3-main-processes-of-geological-change-geology- 91080: yourarticlelibrary.com, Retrieved 25 May, 2019
- Geologic-history-of-Earth: britannica.com, Retrieved 23 April, 2019
- Faults-meaning-classification-and-importance-geology- 91572: yourarticlelibrary.com, Retrieved 16 August, 2019
- Folds-meaning-classification-and-system-geology- 91637: yourarticlelibrary.com, Retrieved 24 July, 2019

Geologic Materials 2

- **Rock**
- **Igneous Rock**
- **Sedimentary Rock**
- **Metamorphic Rock**
- **Magma**
- **Lava**

Geologic materials refer to the various kinds of rock and unlithified material which are found on the surface of the Earth. The major types of rocks are igneous rocks, sedimentary rocks and metamorphic rocks. The topics elaborated in this chapter will help in gaining a better perspective about these geologic materials.

Geologic materials are the rocks and sediments that make up the land where we live. The characteristics of geologic materials reflect the processes that form them and the environments in which they form.

Electrical conductivity (or resistivity) is a bulk property of material describing how well that material allows electric currents to flow through it. Consider current flowing through the unit cube of material shown to the right:

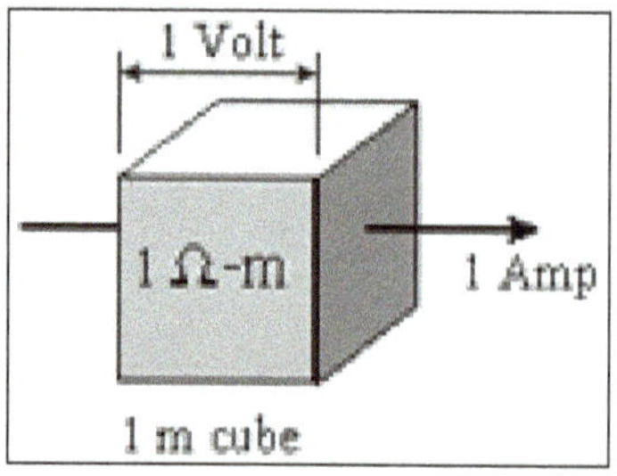

- Resistance is simply the measured voltage over the measured (known) current (which is Ohm's Law). Resistance will change if the measurement geometry or if the volume of material changes. Therefore, it is NOT a physical property.
- Resistivity is basicaly the resistance per unit volume. It is defined as the voltage measured

across a unit cube's length (Volts per metre, or V/m) divided by the current flowing through the unit cube's cross-sectional area (Amps per metre squared, or A/m^2). This results in units of Ohm-m²/m or Ohm-m. The greek symbol rho, ρ, is often used to represent resistivity.

- Conductivity, often represented using sigma, σ, is the inverse of resistivity: $\sigma = 1/\rho$. Conductivity is given in units of Siemens per metre, or S/m. Millisiemens per metre (mS/m) are often used; 1000 mS/m = 1 S/m. So 1 mS/m = 1000 Ohm-m.

The electrical conductivity of Earth's materials varies over many orders of magnitude. It depends upon many factors, including rock type, porosity, connectivity of pores, nature of the fluid, and metallic content of the solid matrix. A very rough indication of the range of conductivity for rocks and minerals is in the figure here.

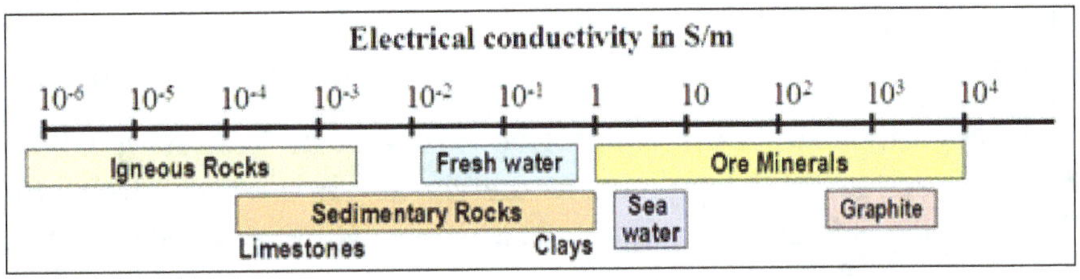

Soils and rocks are composed mostly of silicate minerals, which are essentially insulators. Exceptions include magnetite, specular hematite, carbon, graphite, pyrite, and pyrrhotite. Therefore, conduction is largely electrolytic, so conductivity depends mainly upon:

- Porosity,
- Hydraulic permeability, which describes how pores are interconnected,
- Moisture content,
- Concentration of dissolved electrolytes,
- Temperature and phase of pore fluid,
- Amount and composition of colloids (clay content).

Electrical Conductivity of Metallic Minerals

Metallic ore minerals are relatively uncommon compared to other crustal materials. However, they are often the target of mineral exploration. Even in small quantities, they can significantly affect the bulk resistivity of geologic materials. Most metallic ore minerals are electronic semiconductors. Their resistivities are lower than those of metals and highly variable because the inclusion of impurity ions into a particular metallic mineral has a big effect on the resistivity. For example, pure pyrite has a resistivity of about 3×10^{-5} Ohm-m, but mixing in minor amounts of copper can increase the resistivity six orders of magnitude to 10 Ohm-m. Conductivity properties of some important minerals can be summarized as follows:

- Pyrrhotite (FeS) is a consistently highly conductive mineral.
- Graphite (C) is a true conductor like a metal (i.e. not a semiconductor like ore minerals),

and it is very conductive, even in very low concentrations. It is also chargeable and it is notoriously difficult to distinguish from metallic ore minerals.

- Pyrite (FeS_2) is the most common metallic sulfide and has the most variable conductivity. Its conductivity is generally higher than porous rocks.
- Galena (PbS) and magnetite (Fe_3O_4) are conductive as minerals, but much less conductive as ore because of their loose crystal structures.
- Other conductive minerals include Bornite ($CuFeS_4$), chalcocite (Cu_2S), covellite (CuS), ilmenite ($FeTiO_3$), molybdenite (MoS_2), and the manganese minerals holandite and pyrolusite.
- Hematite and zincblende are usually nearly insulators.

Although metallic minerals (particularly sulfides) may be conductive, there are at least two reasons why ore-grade deposits of these minerals may not be as conductive as expected:

- Sulfide deposits can be either disseminated or massive. In the first type, the mineral occurs as fine particles dispersed throughout the matrix, and in the second, the mineral occurs in a more homogeneous form. Disseminated sulfides may be resistive or conductive, whereas massive sulfides are likely to be conductive.
- Chemical and thermal alteration can convert metallic minerals into oxides or other forms that are not as conductive as the original minerals.

Electrical Properties of Rocks

Of all the geophysical properties of rocks, electrical resistivity is, by far, the most variable. Values ranging as much as 10 orders of magnitude may be encountered, and even individual rock types can vary by several orders of magnitude. The next figure is a representative chart illustrating very generally how the resistivities of important rock groups compare to each other. This type of figure is given in most texts on applied geophysics.

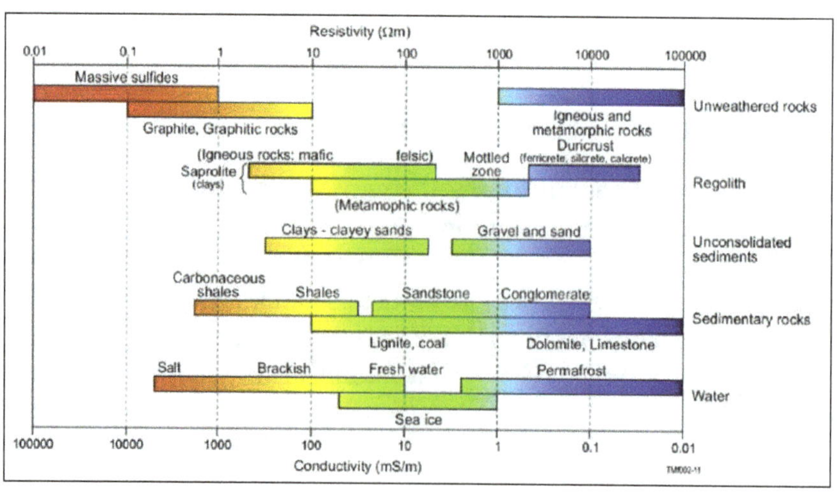

Soils and rocks are composed mostly of silicate minerals, which are essentially insulators, meaning that they have low electrical conductivity. The most common exceptions include magnetite, specular hematite, carbon, graphite, pyrite, and pyrrhotite. Therefore, conduction is largely electrolytic, and conductivity depends mainly upon:

- Porosity,
- Hydraulic permeability, which describes how pores are interconnected,
- Moisture content,
- Concentration of dissolved electrolytes,
- Temperature and phase of pore fluid,
- Amount and composition of colloids (clay content).

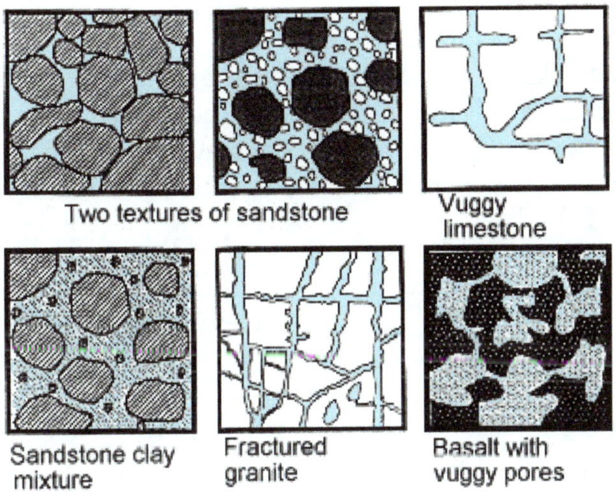

Pore space and pore geometry are the most significant factors. Porosity exists mostly in joints, fractures, vugs (dissolved pockets in limestones and dolomites), and intergranular voids in sedimentary rocks. The figure above and tables below give some idea of the complexity, and range of porosities possible.

Table: Representative porosity ranges for sedimentary materials.

Material	Porosity, %
Soils	50-60
Clay	45-55
Silt	40-50
Medium to coarse mixed sand	35-40
Gravel	30-40
Gravel and sand	20-35
Sandstone	10-20
Shale	1-10

Table: Porosity of soils and rocks.

Rock or Formation	Porosity	Ratio $\frac{\rho_a}{\rho_i}$
Igneous and metamorphic rocks	$\frac{1}{2}$-2	100
Dense limestones and sandstones	3-4	50-100
Clays and sands in general	8-15	20-40
Porous clays,sands,sandstone, cellular limestones, And dolomites	15-40	3-20
Marl, loess, clay, and sandy soil	40-75	1.5-4
Peat,diatomaceous Diatomaceous earth	80-9	1.0-1.5

The "Ratio" column is bulk resistivity divided by electrolyte resistivity.

Vuggy porosity (composed of larger discrete voids) may have very low permeability, making for low resistivity when measured using galvanic (DC current) techniques. However, inductively measured resistivity (using electromagentic induction methods) may be higher because currents induced by oscillating electromagnetic fields do not have to flow over large distances.

Table: Normal ranges in porosity for rocks after Keller and Frisch knechto.

Rock type	Intergranular Porosity (%)	Joint Porosity (%)	Vugular porosity (%)
Paleozoic sandstones and shale	5-30	0-1	0
Paleozoic limestones	2-10	0-2	0
Paleozoic clastic volcanics	5-30	0-2	0
Post Paleozoic sandstones and shale	10-40	0	0
Post-Paleozoic limestones	4-20	0-2	0
Post-Paleozoic clastic volcanics	10-60 max	0	0
Precambrian sediments and low-rank metamorphosed sediments	1-8	0-2	0
Precambrian igneous rocks and high-rank metamorphic rocks	0-2	0-2	0
More recent igneous rocks	0-10	0-2	0
Vugular porosity accounts for and appreciable total porosity only in rare cases.			

Resistivity may be anisotropic in layered rocks, especially for shales where the coefficient of anisotropy (ratio of transverse resistivity to longitudinal resistivity) can be as high as 4.

Much of our understanding about resistivity of porous rocks comes from the oil/gas well-logging industry. The effect of fluids other than water, Archie's law, formation factor, etc. are detailed in the next few sections.

Electrolytes in the Ground

Conductivity of fluids depends upon quantity and mobility (velocity) of charge carriers. Mobility depends on viscosity of fluid (hence temperature) and diameter of charge carriers. Temperature dependence is significant. For sodium chloride solutions, change of conductivity is roughly 2.2% per degree C. So a change of 40oC doubles the conductivity. In the illustration showing conductivities of waters in the Great Lakes (below), compare conductivities in igneous (western) versus sedimentary (eastern) regions, and note the temperature dependence of these lake waters.

Typical conductivities of electrolytes, and examples from the Great Lakes.

Natural source	mS/m
Meteoric waters (from precipitation)	1 to 30
Surface waters (lakes & rivers)	0.3 for very pure waters 10,000 for salt lakes 2 to 30 in igneous regions 10 to 100 in sedimentary regions
Soil waters	Up to 10,000 average around 10
Ground water	6 to 30 in igneous regions 1,000 in sedimentary regions
Mine waters (copper, zinc, etc. i.e. sulfates)	not usually less than 3,000
Lake Superior is the westernmost lake and, therefore, in an igneous region, while Lake Ontario is the easternmost or sedimentary region. This may contribute to the generally more conductive waters of the eastern lakes.	

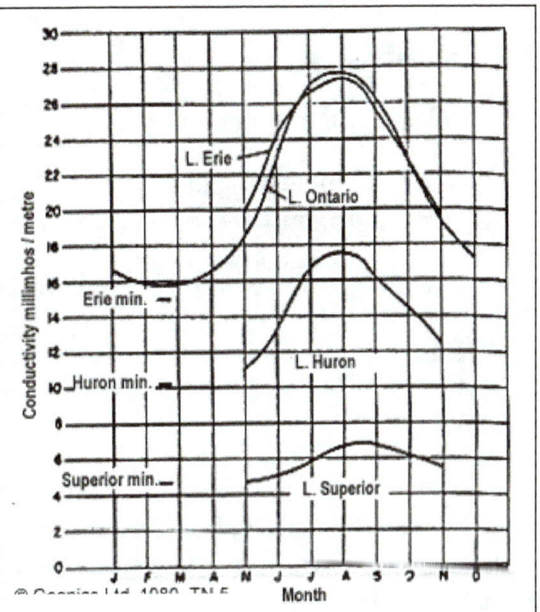

The dependence of fluid conductivity on salinity (concentration of ions) for a variety of electrolytes is illustrated to the right. Tap water is usually a minimum of around 100 Ohm-m with a salinity of about 40 ppm, and sea water is roughly 0.3 Ohm-m with a salinity of 30,000 ppm. Compare these values to those of lake water shown above.

Fluid resistivity depends also upon temperature because the mobility of the ions in solution increases with temperature. This behaviour is opposite to that of metallic conductors (which involve electronic conduction rather than ionic conduction), in which the resistivity increases with temperature.

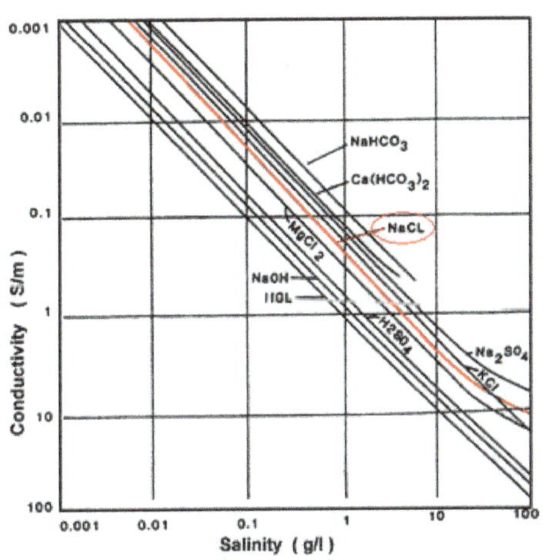

Adapted from Keller and Frischknecht, 1996.

An approximate formula for the resistivity as a function of temperature is:

$$R(t) = \frac{R_{18^o c}}{1 + a(t - 18)}$$

Where, R is resistivity, t is temperature, a is approximately 0.025, and R_{18C} is resistivity at room temperature (18 degrees C). Recall that resistivity=1/conductivity.

Effect of Porosity

Saturated clean (containing no clay) soils or rocks:

Archie's empirical formula relates porosity and water conductivity to bulk conductivity for a variety of consolidated rocks, as well as for unconsolidated materials. Referring to the figure below, this relation is $\frac{\sigma_x}{\sigma_1} = n^m$, where σ_x is bulk conductivity, σ_1 is connate water conductivity, n is porosity (fraction of total volume), and m is a constant. A value of m of around 1.2 is appropriate for spherical particles, and a value near 1.85 is used for platy particles. This parameter is typically 1.4-1.6 for sands.

Another way of expressing Archie's relation, more commonly used by the oil/gas well-logging industry, is: $F=1/\phi^m$, where F, the "formation factor," is $F = R_o / R_w$, R_o is bulk resistivity if pore space is filled 100% with brine (connate water), R_w is resistivity of the connate water itself, and ϕ is porosity. As always, don't be confused by the use of conductivity or resistivity - they are simply reciprocals of each other. It is easy to use a spreadsheet to explore how Archie's equations dictate how porosity and resistivity are related in different materials.

Unsaturated Clean (Containing no Clay) Soils

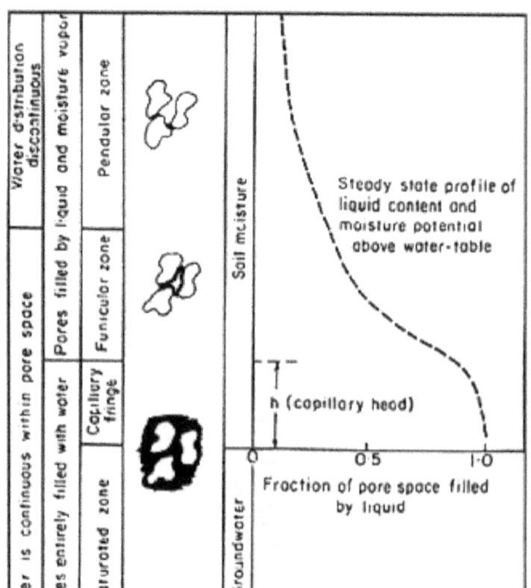

In the funicular zone of soils, moisture does not completely fill pore spaces, but there are still conduction paths. A law similar to Archie's can be used, where n is now the fraction of pore volume filled with electrolyte instead of porosity, and m = 2. Using this, conductivity appears to be very small for low moisture contents.

However, the "wetting" of material is critical in affecting conductivity, and slightly wet materials are much more conductive than dry materials. The relation shown below is similar to Archie's, and gives water saturation in clean (containing no clay) formations, where ϕ is porosity, w is resistivity of water, t is total resistivity, and a and m are both empirical. This relation is hard to use, and definitely wrong in "dirty" (clayey) material.

$$S_W = \frac{volume\ of\ water\ in\ proes}{total\ volume\ of\ pores} = \left(\frac{a\rho_w}{\phi^m \rho_t}\right)^{(1/m)}$$

Therefore, water saturation may be estimated if:

1. Electrical methods can be used to find formation resistivity,
2. If connate water can be tested, and
3. If porosity can be estimated.

This is similar to finding water saturation, Sw, when a portion of the pore space is filled with oil or gas, as is done all the time using well-logging data in hydrocarbon reservoirs.

Resistivity of Soils

The electrical conductivity of soils is rather complicated, with many factors affecting the bulk properties. The following material is not covered in most texts on applied geophysics, but it is important because soils are usually (except for borehole work) the closest materials to the survey electrodes and, therefore, soils have a significant effect on results. Porosity ranges from 20 to 70% for most unconsolidated materials (i.e. for soils). However, it is not common to have a large range of porosities in one situation. As noted above, porosity is the primary property related to resistivity, hence the difficulty in discriminating between sand and gravel, as long as porosity is similar.

Effect of Freezing on Conductivity of Soils

Reducing temperature reduces electrolytic activity and, hence, conductivity. Upon freezing, conductivity of water becomes that of ice, which is very low. However, freezing is rarely simple. Fresh water freezes at a higher temperature than saline water; therefore, solutes tend to become consentrated in a zone of unfrozen saline water adjacent to particles. Also, the electric field of cations adsorbed onto particles locally orients water molecules near the particle, preventing them from freezing.

The net effect is of a steady (slight) decrease in conductivity as temperatures approach freezing, then a leveling off through 0 degrees, and a further decrease below freezing. As expected, this is most noticeable in clayey material.

Colloidal Conductivity - Conductivity due to Clay

Complexity and variety of soil types is illustrated in the ternary diagram below left. It does not take much clay to change electrical properties of soils. Any fine grained mineral exhibits a certain Cation Exchange Capacity (CEC). That is, charges (cations) can be adsorbed (attached to the surface) onto the slightly negatively charged surface, and these can subsequently be exchanged or dissolved.

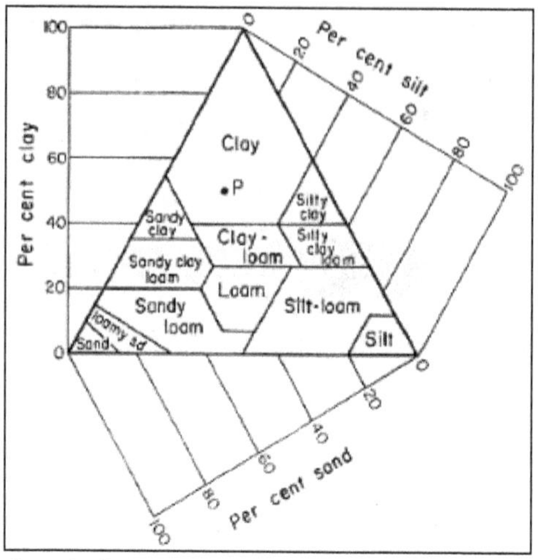

US Department of Agriculture textural classification triangle (ternary diagram).
Point P is a clay (soil) with 50% clay, 20% silt, 30% sand.

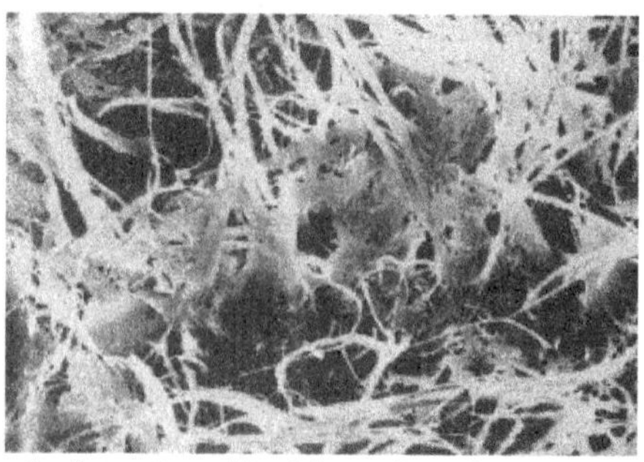

Illite (a clay mineral) with total surface area to volume ratio of 100m2/cm³.

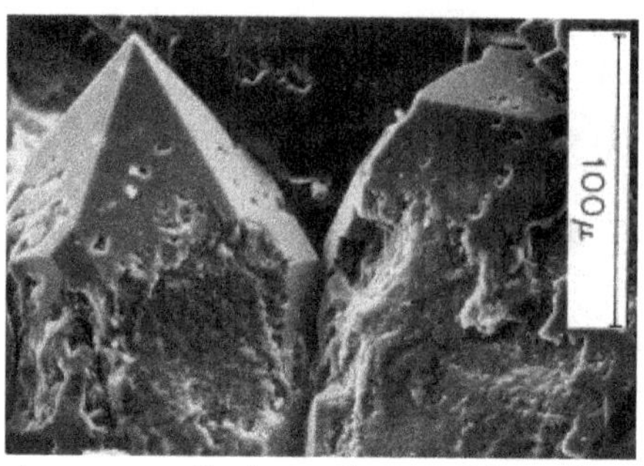

Quartz overgrowths in sandstone with total surface area to volume ratio of 0.1m²/cm³.

Clay has a huge surface-area to volume ratio, so it has a much higher exchange capacity, especially vermiculite and montmorillonite. Therefore, clays can dramatically increase the conductivity of connate water, especially of "fresher" waters. (Saline waters may not have much more capacity to absorb extra electrolytes.

Aspects of Soil Formation affecting Electrical Properties of Soils

It is worth discussing the formation of soils in order to gain a better appreciation for what is involved when predicting electrical properties of near-surface materials, and when interpreting shallow geophysical surveys. This discussion does not replace a course on soil science, but some issues that affect electrical resistivity should become more clear. Generally, electrical properties are affected by varying clay content or ionic type and concentration in water. The following is an outline of how these factors evolve in soils.

Weathering involves mechanical, chemical and biological processes that convert surficial materials to humus and clay and fine-grained sediments. In the presence of water and CO_2, rocks are broken down into ions (often dissolved and removed by drainage), clay minerals are formed, water is used up (becomes part of clay compounds), and solutions become more basic (less acidic). The process is self-perpetuating, since a thin layer of soil will cause the relevant processes to occur more rapidly at the rock surface; i.e. the layer retains water and CO_2, which produces weak carbonic acid, which combines with rock components to form clays.

Speed of weathering is a function of temperature, vegetative growth, and availability of moisture. Therefore, tropical soils tend to be thick. Well-drained soils tend to be devoid of unstable minerals (i.e. electrolytes), and dry soils tend to be saline (therefore, conductive).

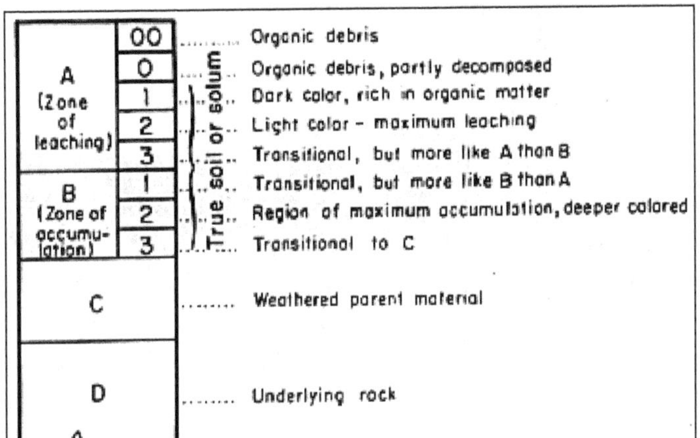

Normal or zonal soil profile

The figure is a typical soil profile:

- The A zone is typically intensely weathered, permeable and devoid of solutes.
- The B zone is generally much tighter (less porosity), richer in clays (both local and transported from A), and controls how much water gets to the C zone and underlying rock layers.
- The C zone is generally permeable, and consists of less-weathered parent material.
- The D zone is often treated as non-porous and impermeable.

Soil moisture is affected by several factors. Refer to the figure above illustrating typical distribution of moisture in soils above the water table.

- In the pendular zone, water exists in isolated rings around tight spots. In the funicular zone, there is a thin layer of water over the surface area. Thickness of this depends on capillary forces.
- If there is fine grained material over coarse layering, the fine grained region may have funicular water, while the coarse layer may have pendular water, and may, therefore, have a lower conductivity.
- Behaviour of the water table depends upon many things including permeability (which ranges by a factor of 1010), and regional humidity, as sketched to the right. These factors can cause many kinds of water table configurations, some of which may be rather counter-intuitive.

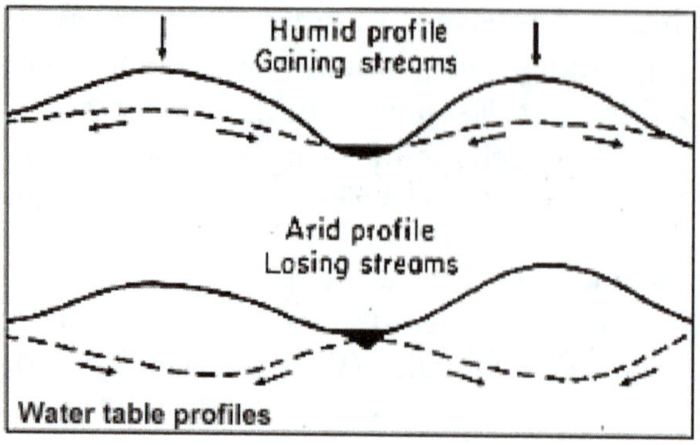

Anisotropic Ground

Anisotropy means "depending upon direction" Structural anisotropy (for example, layering or fracturing) may cause the electrical properties of the ground to be anisotropic. This means that measured apparent resistivity will depend upon the direction of the measurement system, as in the adjacent figure. Anisotropy may be very interesting; for example, preferential directions of fluid flow may well be determined by measuring how resistivity varies as a function of the orientation of the measurement electrodes (eg. north-south versus east-west). However, if anisotropy exists but is ignored, then true ground resistivities that are interpreted from measured apparent resistivities may not be correct.

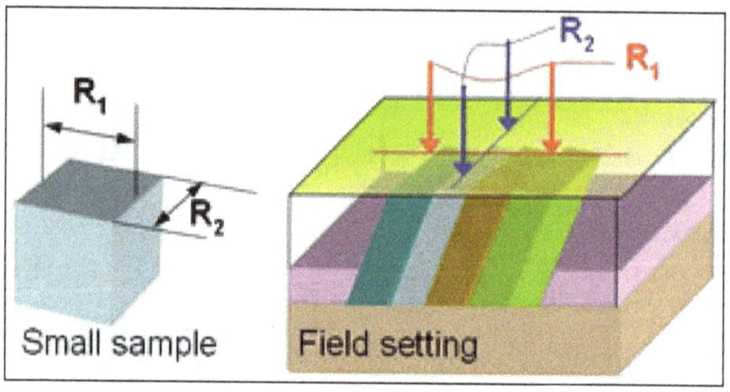

For anisotropic materials, R_1 not equal R_2.

Vertically Anisotropic Ground

For normal resistivity surveys carried out at the surface, there is no way to tell the difference between resistivity measured verically and resistivity measured horizontally. Therefore, vertical anisotropy is undetectable at the surface. If such anisotropy exists, depth estimates will be in error by a factor of λ, the coefficient of anisotropy (defined as the square root of the ratio of vertical to horizontal resistivities).

Horizontally Anisotropic Ground

Horizontal anisotropy means that resistivity measured with electrodes oriented in one direction will be different than that measured using the same array oriented in a perpendicular direction. In general, "transverse resistivity" (resistivity measured perpendicular to the bedding plane) will be greater than "longitudinal resistivity" (measured parallel to the bedding plane).

A Counter-intuitive Effect

It should be noted that the effect of steeply dipping beds on surface resistivity measurements is not as might be first expected. If anisotropy is steeply dipping (and there is no overburden), one might expect that measured resistivity would be lowest parallel to strike, since current tends to flow along paths of least resistance. In fact, measured resistivity is highest along strike because of the increased current density parallel to the survey. Apparent resistivity calculations assume uniform current density in three dimensions; so when current density is higher than it would be in a half space, measured potential difference is higher for the given current source, resulting in a higher apparent resistivity. Therefore, resistivities measured with arrays placed along strike are over-estimated and resistivities measured perpendicular to strike are under-estimated.

Why Anisotropy Occurs

Structural anisotropy (for example, layering or fracturing) causes the simple form of Ohm's law to break down because current flow is not necessarily parallel to the forcing electric field. Instead of simply writing $J_i = \sigma E = -\sigma \Delta V$, we have to write,

$$J_i = -\sigma_{ik} \frac{\partial V}{\partial x_k} i, k = 1,2,3.$$

In homogeneous ground with single current and potential electrodes, the expression for V (voltage) in terms of resistivity and distance from the current source is $V = -I\rho / 2\pi r$. In anisotropic ground, there are both horizontal and a vertical resistivities. The expression for voltage in terms of the horizontally and vertically oriented resistivities and distance is $V = -I\sqrt{\rho_h \rho_v} / 2\pi r = -I\rho_h \lambda / 2\pi r$, where $\lambda = (\rho_v \rho_h)^{1/2}$ is called the coefficient of anisotropy.

Archie's Equation and Water Saturation

G.E. Archie's work relating permeability to porosity resulted in empirical relationships between resistivities, fluid types and porosity. In undisturbed formations, Archie's (empirical) law states that the ratio of formation resistivity to connate water resistivity (R_o/R_w) is constant. That is:

$$F = R_o/R_w$$

where,

F is called the formation factor,

R_o is bulk resistivity if pore space is filled 100% with brine (connate water),

R_w is resistivity of the connate water itself. This relation applies best for clean (clay free) rocks of constant porosity when R_w<1 Ohm-m, at 25C.

This is useful, except that Ro is not really measurable in the field. But the relation can be made useful because Archie also found that this resistivity ratio (i.e. F) changes consistently as porosity changes.

Formation factor and porosity are usually related via $F = 1/\phi^m$.

where,

ϕ is porosity (a value between 0 and 1).

m is the cementation exponent and is usually 1.5-1.8 in sandstones and 2.0 in limestones, dolomites and tight consolidated sandstones.

Alternative relations are:

$F = 0.62/\phi^{2.15m}$ (Humble relation, for many granular rocks).

$F = 1/\phi^{(1.87+0.019/)}$ (Shell relation, for low-porosity carbonates).

Finally, water saturation (S_W) and formation resistivity were also found to be related. Water saturation is the percentage of the pore space filled with water, which is, of course, important when evaluating a potential hydrocarbon reservoir. $S_w = 1$ means all fluids are water, and $S_w = 0.1$ means 10% of fluids are water, implying 90% are non-conductive or oil/gas. The relation is:

$$S_W = (R_0 / R_T)^{1/n}$$

where,

- n is a saturation exponent, usually close to 2.0. (n is not porosity in this relation).
- R_o = rock resistivity with 100% brine for fluid.
- R_T = true resistivity (with hydrocarbons and water in pore space).

Now the Archie equation relating quantities of interest to measurable quantities can be derived based upon equations:

$$S_w = (FR_W / R_T)^{1/n}$$

This relation is commonly used for oil/gas reservoir characterization. Three parameters must be measured: (i) porosity, (ii) resistivity of the undisturbed formation, and (iii) resistivity of connate water. Symbols are, R_T and R_W, respectively. R_W can be obtained from:

- Water catalogues of the area you're working in;
- Laboratory analysis of samples extracted from the formation rocks;
- Obtainable from the spontaneous potential or SP log;
- "Apparent" water resistivity from $R_t = F * R_{wa}$ when $S_w = 1$, e.g. in "clean water sands;"
- By relating invaded & uninvaded resistivity in clean zones;
- From resistivity-porosity crossplots;
- From R_{wa}-*SP* crossplots.

R_T is true resistivity of undisturbed formation. Of course, drilling disturbs the formation, so choosing the right tool and interpreting various resistivities is important. Methods must account for borehole environment, invasion, effects of adjacent beds, and technical aspects of tools themselves.

In the oil/gas context, porosity, ϕ, is obtained from the porosity well logging tools, or possibly from resistivity, where S_w is known to be 100%. Porosity logging tools are also used in hydrogeology and engineering situations, though this is not routine as in the oil/gas industry.

Rock

Rock, in geology, is a naturally naturally occurring and coherent aggregate of one or more minerals. Such aggregates constitute the basic unit of which the solid Earth is comprised and typically form recognizable and mappable volumes. Rocks are commonly divided into three major classes according to the processes that resulted in their formation. These classes are (1) igneous rocks, which have solidified from molten material called magma; (2) sedimentary rocks, those consisting of fragments derived from preexisting rocks or of materials precipitated from solutions; and (3) metamorphic rocks, which have been derived from either igneous or sedimentary rocks under conditions that caused changes in mineralogical composition, texture, and internal structure. These three classes, in turn, are subdivided into numerous groups and types on the basis of various factors, the most important of which are chemical, mineralogical, and textural attributes.

Rock Types

Igneous rocks are those that solidify from magma, a molten mixture of rock-forming minerals and usually volatiles such as gases and steam. Since their constituent minerals are crystallized from molten material, igneous rocks are formed at high temperatures. They originate from processes deep within the Earth—typically at depths of about 50 to 200 kilometres (30 to 120 miles)—in the mid- to lower-crust or in the upper mantle. Igneous rocks are subdivided into two categories: intrusive (emplaced in the crust), and extrusive (extruded onto the surface of the land or ocean bottom), in which case the cooling molten material is called lava.

Sedimentary rocks are those that are deposited and lithified (compacted and cemented together) at the Earth's surface, with the assistance of running water, wind, ice, or living organisms. Most are deposited from the land surface to the bottoms of lakes, rivers, and oceans. Sedimentary rocks are generally stratified—i.e., they have layering. Layers may be distinguished by differences in colour, particle size, type of cement, or internal arrangement.

Metamorphic rocks are those formed by changes in preexisting rocks under the influence of high temperature, pressure, and chemically active solutions. The changes can be chemical (compositional) and physical (textural) in character. Metamorphic rocks are often formed by processes deep within the Earth that produce new minerals, textures, and crystal structures. The recrystallization that takes place does so essentially in the solid state, rather than by complete remelting, and can be aided by ductile deformation and the presence of interstitial fluids such as water. Metamorphism often produces apparent layering, or banding, because of the segregation of minerals into separate bands. Metamorphic processes can also occur at the Earth's surface due to meteorite impact events and pyrometamorphism taking place near burning coal seams ignited by lightning strikes.

Rock Cycle

Geologic materials—mineral crystals and their host rock types—are cycled through various forms. The process depends on temperature, pressure, time, and changes in environmental conditions in the Earth's crust and at its surface. The rock cycle illustrated in figure reflects the basic relationships among igneous, metamorphic, and sedimentary rocks. Erosion includes weathering (the physical and chemical breakdown of minerals) and transportation to a site of deposition. Diagenesis is, as previously explained, the process of forming sedimentary rock by compaction and natural cementation of grains, or crystallization from water or solutions, or recrystallization. The conversion of sediment to rock is termed lithification.

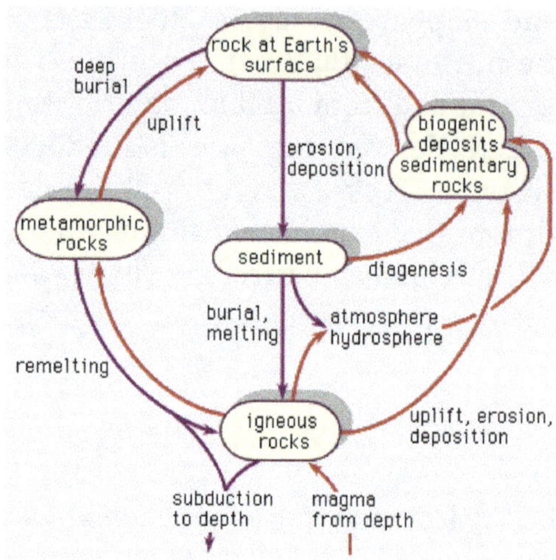

Abundance of Rock Types

An estimate of the distribution of rock types in large structural units of the terrestrial crust is given in the table.

Table: Distribution of Rock Types in Large Structural Units of the Earth's Crust.

Crustal layers	Average thickness (in kilometres)	Volume (in cubic kilometres)	Mass (10^{24} grams)	Rock types and abundances (percent volume of layer)
Continental platform Sedimentary	1.8	135	0.35	Sands, 23.6; clays, 49.5; carbonates, 21; evaporites, 2; basalts, 3.9
Continental geosynclinal folded belts Sedimentary	10.0	365	0.94	Sands, 18.7; clays and shales, 39.4; carbonates, 16.3; evaporites, 0.3; basalts, 12.6; andesites, 102; rhyolites, 2.5
Subcontinental, shelf and slope Sedimentary	2.9	190	0.48	Similar to above groups
Continental and subcontinental Granitic (continental) (subcontinental)	20.1 9.1	3,590	9.81	Granites, 18.1; granodiorites, 19.9; syenites, 0.3; gabbro, 3.7; peridotites, 0.1; gneisses, 37.6; schists, 9.0; marble, 1.5; amphibolite, 9.8

Basaltic (continental) (subcontinental)	20.1 11.7	3,760	10.91	Acid igneous and metamorphic rocks, 50; basic igneous and metamorphic rocks, 50
Oceanic Layer 1 (sedimentary)	0.4	120	0.19	(percent by area) terrigenous, 7.3; calcareous, 41.5; siliceous, 17; red clays, 31.2
Layer 2	0.6 0.6	175 175	0.44 0.52	(percent by volume) sediments, 50 (percent by volume) basalts, 50
Layer 3	5.7	1,700	4.92	(percent by volume) oceanic tholeiitic basalts, 99; alkaline differentiates, 1

Table: Abundances of Main Rock Types and Minerals in the Earth's Crust.

Rocks	Percent volume of crust	Minerals	Percent volume of crust
Sedimentary		Constituents of sedimentary rocks	
Sands	1.7	Quartz	12
Clays and shales	4.2	Alkali feldspar	12
Carbonates (including salt)	2.0	Plagioclase feldspar	39
	7.9	Micas	5
Igneous		Amphiboles	5
Granite	10.4		73
Granodiorite, diorite	11.2	Constituents of igneous rocks	
Syenite	0.4	Pyroxene	11.0
Basalt, gabbro, amphibolite, eclogite	42.5	Olivine	3.0
Erunite, periodotite	0.2	Clay minerals (and chlorite)	4.6
	64.7	Calcite	1.5
		Dolomite	0.5
Metamorphic		(Titano) magnetite	1.5
Gnei ss	21.4	Others (garnet, kyanite, apatite, etc.)	4.9
Schist	5.1		27.0
Marble	0.9	Totals	
	27.4	Quartz and feldspar	63
Totals		Pyroxene and olivine	14
Sedimentary	7.9	Hydrated silicates	14.6
Igneous	64.7	Carbonates	2.0

Metamorphic	27.4		100.0
	100.0	Others	6.4

Texture

The texture of a rock is the size, shape, and arrangement of the grains (for sedimentary rocks) or crystals (for igneous and metamorphic rocks). Also of importance are the rock's extent of homogeneity (i.e., uniformity of composition throughout) and the degree of isotropy. The latter is the extent to which the bulk structure and composition are the same in all directions in the rock.

Rocks have many different textures. Layered sandstone produces a gritty texture, whereas coquina may be rough with cemented shells occasionally producing a sharp edge. Likewise, breccia, which contains pieces of other rocks that have been cemented together, and porphyry, which contains interlocking mineral crystals, tend to be rough. In contrast, obsidian tends to have a smooth glassy feel, whereas serpentine may feel platy or fibrous, and talc schist often feels greasy. On the other hand, the texture of gneiss is often described by its distinct banding.

Analysis of texture can yield information about the rock's source material, conditions and environment of deposition (for sedimentary rock) or crystallization and recrystallization (for igneous and metamorphic rock, respectively), and subsequent geologic history and change.

Classification by Grain or Crystal Size

The common textural terms used for rock types with respect to the size of the grains or crystals, are given in the Table. The particle-size categories are derived from the Udden-Wentworth scale developed for sediment. For igneous and metamorphic rocks, the terms are generally used as modifiers—e.g., medium-grained granite. Aphanitic is a descriptive term for small crystals, and phaneritic for larger ones. Very coarse crystals (those larger than 3 centimetres, or 1.2 inches) are termed pegmatitic.

For sedimentary rocks, the broad categories of sediment size are coarse (greater than 2 millimetres, or 0.08 inch), medium (between 2 and 1/16 millimetres), and fine (under 1/16 millimetre). The latter includes silt and clay, which both have a size indistinguishable by the human eye and are also termed dust. Most shales (the lithified version of clay) contain some silt. Pyroclastic rocks are those formed from clastic material ejected from volcanoes. Blocks are fragments broken from solid rock, while bombs are molten when ejected.

Table: Common textural terms for rocks.

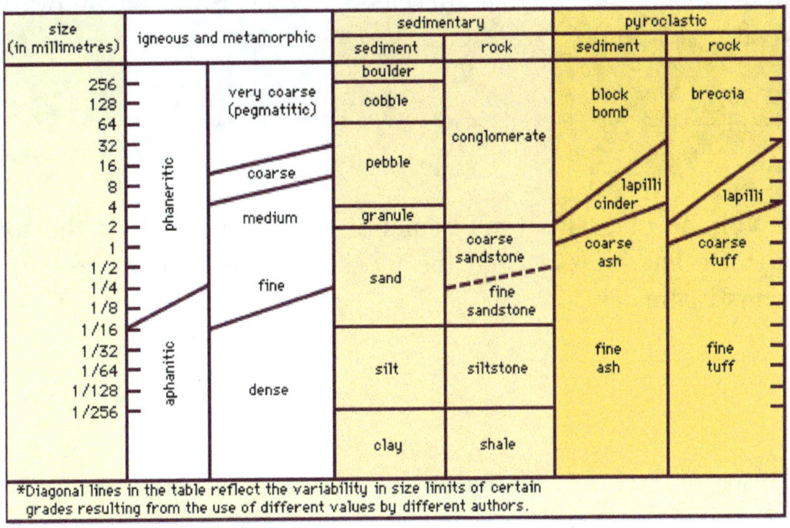

Porosity

The term rock refers to the bulk volume of the material, including the grains or crystals as well as the contained void space. The volumetric portion of bulk rock that is not occupied by grains, crystals, or natural cementing material is termed porosity. That is to say, porosity is the ratio of void volume to the bulk volume (grains plus void space). This void space consists of pore space between grains or crystals, in addition to crack space. In sedimentary rocks, the amount of pore space depends on the degree of compaction of the sediment (with compaction generally increasing with depth of burial), on the packing arrangement and shape of grains, on the amount of cementation, and on the degree of sorting. Typical cements are siliceous, calcareous or carbonate, or iron-bearing minerals.

Sorting is the tendency of sedimentary rocks to have grains that are similarly sized—i.e., to have a narrow range of sizes. Poorly sorted sediment displays a wide range of grain sizes and hence has decreased porosity. Well-sorted indicates a grain size distribution that is fairly uniform. Depending on the type of close-packing of the grains, porosity can be substantial. It should be noted that in engineering usage—e.g., geotechnical or civil engineering—the terminology is phrased oppositely and is referred to as grading. A well-graded sediment is a (geologically) poorly sorted one, and a poorly graded sediment is a well-sorted one.

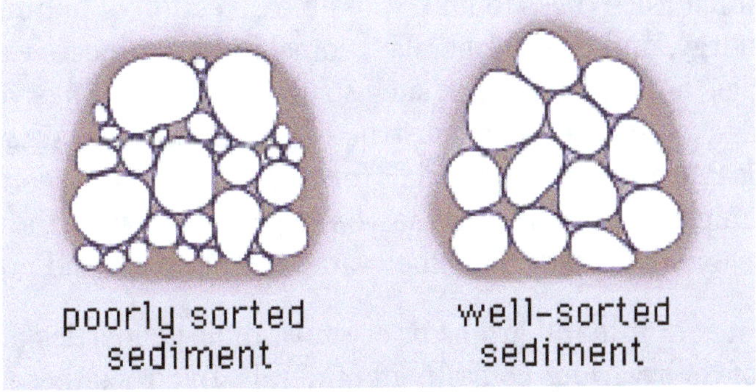

Sorting.

Total porosity encompasses all the void space, including those pores that are interconnected to the surface of the sample as well as those that are sealed off by natural cement or other obstructions. Thus the total porosity (ϕ_T) is:

$$\varphi\mathrm{T} = 100\left(1 - \frac{\mathrm{Vol_G}}{\mathrm{Vol_B}}\right)\%$$

where $\mathrm{Vol_G}$ is the volume of grains (and cement, if any) and $\mathrm{Vol_B}$ is the total bulk volume. Alternatively, one can calculate (ϕ_T) from the measured densities of the bulk rock and of the (mono)mineralic constituent. Thus,

$$\phi_\mathrm{T} = 100\left(1 - \frac{\mathrm{Vol_G}}{\mathrm{Vol_B}}\right)\%$$

where ρ_B is the density of the bulk rock and ρ_G is the density of the grains (i.e., the mineral, if the composition is monomineralogic and homogeneous). For example, if a sandstone has a ρ_B of 2.38 grams per cubic centimetre (g/cm³) and is composed of quartz (SiO_2) grains having ρ_G of 2.65 g/cm³, the total porosity is

$$\varphi T = 100\left(1 - \frac{2.38}{2.65}\right) = 10.2\%,$$

Apparent (effective, or net) porosity is the proportion of void space that excludes the sealed-off pores. It thus measures the pore volume that is effectively interconnected and accessible to the surface of the sample, which is important when considering the storage and movement of subsurface fluids such as petroleum, groundwater, or contaminated fluids.

Physical Properties

Physical properties of rocks are of interest and utility in many fields of work, including geology, petrophysics, geophysics, materials science, geochemistry, and geotechnical engineering. The scale of investigation ranges from the molecular and crystalline up to terrestrial studies of the Earth and other planetary bodies. Geologists are interested in the radioactive age dating of rocks to reconstruct the origin of mineral deposits; seismologists formulate prospective earthquake predictions using premonitory physical or chemical changes; crystallographers study the synthesis of minerals with special optical or physical properties; exploration geophysicists investigate the variation of physical properties of subsurface rocks to make possible detection of natural resources such as oil and gas, geothermal energy, and ores of metals; geotechnical engineers examine the nature and behaviour of the materials on, in, or of which such structures as buildings, dams, tunnels, bridges, and underground storage vaults are to be constructed; solid-state physicists study the magnetic, electrical, and mechanical properties of materials for electronic devices, computer components, or high-performance ceramics; and petroleum reservoir engineers analyze the response measured on well logs or in the processes of deep drilling at elevated temperature and pressure.

Since rocks are aggregates of mineral grains or crystals, their properties are determined in large part by the properties of their various constituent minerals. In a rock these general properties are determined by averaging the relative properties and sometimes orientations of the various grains or crystals. As a result, some properties that are anisotropic (i.e., differ with direction) on a submicroscopic or crystalline scale are fairly isotropic for a large bulk volume of the rock. Many properties are also dependent on grain or crystal size, shape, and packing arrangement, the amount and distribution of void space, the presence of natural cements in sedimentary rocks, the temperature and pressure, and the type and amount of contained fluids (e.g., water, petroleum, gases). Because many rocks exhibit a considerable range in these factors, the assignment of representative values for a particular property is often done using a statistical variation.

Some properties can vary considerably, depending on whether measured in situ (in place in the subsurface) or in the laboratory under simulated conditions. Electrical resistivity, for example, is highly dependent on the fluid content of the rock in situ and the temperature condition at the particular depth.

Density

Density varies significantly among different rock types because of differences in mineralogy and porosity. Knowledge of the distribution of underground rock densities can assist in interpreting subsurface geologic structure and rock type.

In strict usage, density is defined as the mass of a substance per unit volume; however, in common usage, it is taken to be the weight in air of a unit volume of a sample at a specific temperature. Weight is the force that gravitation exerts on a body (and thus varies with location), whereas mass (a measure of the matter in a body) is a fundamental property and is constant regardless of location. In routine density measurements of rocks, the sample weights are considered to be equivalent to their masses, because the discrepancy between weight and mass would result in less error on the computed density than the experimental errors introduced in the measurement of volume. Thus, density is often determined using weight rather than mass. Density should properly be reported in kilograms per cubic metre (kg/m³), but is still often given in grams per cubic centimetre (g/cm³).

Another property closely related to density is specific gravity. It is defined, as noted above, as the ratio of the weight or mass in air of a unit volume of material at a stated temperature to the weight or mass in air of a unit volume of distilled water at the same temperature. Specific gravity is dimensionless (i.e., has no units).

The bulk density of a rock is $\rho_B = W_G / V_B$, where W_G is the weight of grains (sedimentary rocks) or crystals (igneous and metamorphic rocks) and natural cements, if any, and V_B is the total volume of the grains or crystals plus the void (pore) space. The density can be dry if the pore space is empty, or it can be saturated if the pores are filled with fluid (e.g., water), which is more typical of the subsurface (in situ) situation. If there is pore fluid present,

$$(saturated)\ \rho_B = \frac{W_G + W_{fl'}}{V_B},$$

where, W_{fl} is the weight of pore fluid. In terms of total porosity, saturated density is

$$(saturated)\ \rho_B = \rho_G(1-\varphi_T) + \rho_{fl}\varphi_T,$$

$$\text{and thus } \varphi_T = \frac{\rho_G - \rho_B}{\rho_G - \rho_{fl}},$$

where, ρ_{fl} is the density of the pore fluid. Density measurements for a given specimen involve the determination of any two of the following quantities: pore volume, bulk volume, or grain volume, along with the weight.

A useful way to assess the density of rocks is to make a histogram plot of the statistical range of a set of data. The representative value and its variation can be expressed as follows: (1) mean, the average value, (2) mode, the most common value (i.e., the peak of the distribution curve), (3) median, the value of the middle sample of the data set (i.e., the value at which half of the samples are below and half are above), and (4) standard deviation, a statistical measure of the spread of the data (plus and minus one standard deviation from the mean value includes about two-thirds of the data).

A compilation of dry bulk densities for various rock types found in the upper crust of the Earth is listed in the table. A histogram plot of these data, giving the percent of the samples as a function of density is shown in figure. The parameters given include (1) sample division, the range of density in one data column—e.g., 0.036 g/cm³ for figure, (2) number of samples, and (3) standard deviation. The small inset plot is the percentage of samples (on the vertical axis) that lie within the interval of the "mode - x" to the "mode + x," where x is the horizontal axis.

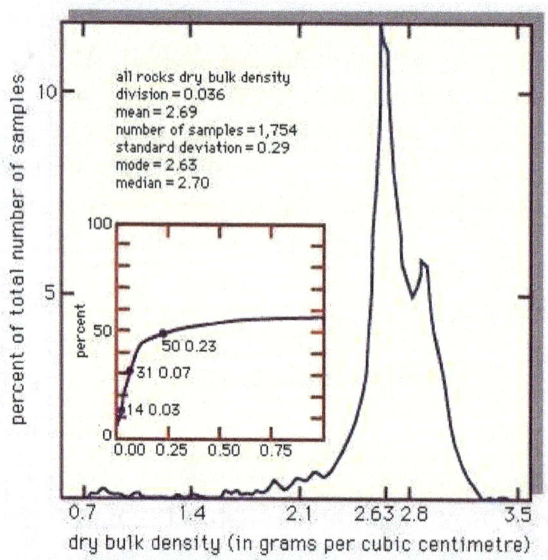

Dry bulk densities (distribution with density) for all rocks.

In figure, the most common (modal) value of the distribution falls at 2.63 g/cm³, roughly the density of quartz, an abundant rock-forming mineral. Few density values for these upper crustal rocks lie above 3.3 g/cm³. A few fall well below the mode, even occasionally under 1 g/cm³. The reason for this is shown in figure, which illustrates the density distributions for granite, basalt, and sandstone. Granite is an intrusive igneous rock with low porosity and a well-defined chemical (mineral) composition; its range of densities is narrow. Basalt is, in most cases, an extrusive igneous rock that can exhibit a large variation in porosity (because entrained gases leave voids called vesicles), and thus some highly porous samples can have low densities. Sandstone is a clastic sedimentary rock that can have a wide range of porosities depending on the degree of sorting, compaction, packing arrangement of grains, and cementation. The bulk density varies accordingly.

The density of clastic sedimentary rocks increases as the rocks are progressively buried. This is because of the increase of overburden pressure, which causes compaction, and the progressive cementation with age. Both compaction and cementation decrease the porosity.

Representative densities for common rock-forming minerals (i.e., ρ_G) and rocks (i.e., ρ_B) are listed in the Minerals and Rocks.

Table: Typical Densities.

material		ρ (g/cm³)	
minerals (approximately pure, no porosity)		ρ_G	
Halite (NaCl)		2.16	
Gypsum ($CaSO_4 \cdot 2H_2O$)		2.32	
Orthoclase feldspar ($KAlSi_3O_8$)		2.56	
Quartz (SiO_2)		2.65	
Calcite ($CaCO_3$)		2.71	
Anhydrite ($CaSO_4$)		2.9	
Olivine $(Mg,Fe)_2SiO_4$		3.3–3.6	
Hematite (Fe_2O_3)		5.26	
rocks (surface T and P)	typical porosity	dry ρ_B	saturated ρ_B
Loess		1.1	1.4–1.6
Silt		1.36	1.4–1.0
Rock salt		2.2	—
Shale	≳8%	1.9–2.4	2.1–2.5
Sandstone	~15%, variable	2.0–2.6	2.2–2.6
Limestone	~5%	1.9–2.6	2.1–2.7
Dolomite		2.5–2.8	2.6–2.8
Granite-diorite		2.6–2.9	
Gabbro-diabase		2.8–3.1	
Dunite-peridotite		3.1–3.4	
Granite; sial crust		2.67	
Basalt; sima crust		2.87	
Top of mantle		3.32	
other		ρ	
Water (0 °C)		0.99987	
20 °C		0.99823	
100 °C		0.95841	
Seawater (0° C)		1.025	
Ice		0.92	
Oil		~0.8–0.9	
Methane gas (at 200 bars, $T \sim 80°$ C)		~0.12	

In table, the bulk densities for sedimentary rocks, which typically have variable porosity, are given as ranges of both dry ρ_B and (water-) saturated ρ_B. The pore-filling fluid is usually briny water, often indicative of the presence of seawater when the rock was being deposited or lithified. It should be noted that the bulk density is less than the grain density of the constituent mineral (or mineral assemblage), depending on the porosity. For example, sandstone (characteristically quartzose) has a typical dry bulk density of 2.0–2.6 g/cm³, with a porosity that can vary from low to more than 30 percent. The density of quartz itself is 2.65 g/cm³. If porosity were zero, the bulk density would equal the grain density.

Saturated bulk density is higher than dry bulk density, owing to the added presence of pore-filling fluid. The Table also lists representative values for density of seawater, oil, and methane gas at a subsurface condition—pressure of 200 bars (one bar = 0.987 atmosphere, or 29.53 inches of mercury) and a temperature of about 80 °C (176 °F).

Mechanical Properties

Stress and Strain

When a stress σ (force per unit area) is applied to a material such as rock, the material experiences a change in dimension, volume, or shape. This change, or deformation, is called strain (ε). Stresses can be axial—e.g., directional tension or simple compression—or shear (tangential), or all-sided (e.g., hydrostatic compression). The terms stress and pressure are sometimes used interchangeably, but often stress refers to directional stress or shear stress and pressure (P) refers to hydrostatic compression. For small stresses, the strain is elastic (recoverable when the stress is removed and linearly proportional to the applied stress). For larger stresses and other conditions, the strain can be inelastic, or permanent.

Elastic Constants

In elastic deformation, there are various constants that relate the magnitude of the strain response to the applied stress. These elastic constants include the following:

1. Young's modulus (E) is the ratio of the applied stress to the fractional extension (or shortening) of the sample length parallel to the tension (or compression). The strain is the linear change in dimension divided by the original length.
2. Shear modulus (μ) is the ratio of the applied stress to the distortion (rotation) of a plane originally perpendicular to the applied shear stress; it is also termed the modulus of rigidity.
3. Bulk modulus (k) is the ratio of the confining pressure to the fractional reduction of volume in response to the applied hydrostatic pressure. The volume strain is the change in volume of the sample divided by the original volume. Bulk modulus is also termed the modulus of incompressibility.
4. Poisson's ratio (σ_p) is the ratio of lateral strain (perpendicular to an applied stress) to the longitudinal strain (parallel to applied stress).

For elastic and isotropic materials, the elastic constants are interrelated. For example,

$$\sigma_p = \frac{E}{2\mu} - 1;$$

$$E = 3k\left(1 - 2\sigma_p\right);$$

and

$$\mu = \frac{3}{2}k\left(\frac{1 - 2\sigma_p}{1 - \sigma_p}\right).$$

The following are the common units of stress:

$$1\,bar = 10^6 \text{ dynes per square centimetre}$$
$$= 10^5 \text{ newtons per square metre, or pascals(pa)}$$
$$= 0.1 \text{ megapascal } (i.e., 0.1 \times 10^6\,Pa).$$

Thus 10 kilobars = 1 gigapascal (i.e., 10^9 Pa).

Rock Mechanics

The study of deformation resulting from the strain of rocks in response to stresses is called rock mechanics. When the scale of the deformation is extended to large geologic structures in the crust of the Earth, the field of study is known as geotectonics.

The mechanisms and character of the deformation of rocks and Earth materials can be investigated through laboratory experiments, development of theoretical models based on the properties of materials, and study of deformed rocks and structures in the field. In the laboratory, one can simulate—either directly or by appropriate scaling of experimental parameters—several conditions. Two types of pressure may be simulated: confining (hydrostatic), due to burial under rock overburden, and internal (pore), due to pressure exerted by pore fluids contained in void space in the rock. Directed applied stress, such as compression, tension, and shear, is studied, as are the effects of increased temperature introduced with depth in the Earth's crust. The effects of the duration of time and the rate of applying stress (i.e., loading) as a function of time are examined. Also, the role of fluids, particularly if they are chemically active, is investigated.

Some simple apparatuses for deforming rocks are designed for biaxial stress application: a directed (uniaxial) compression is applied while a confining pressure is exerted (by pressurized fluid) around the cylindrical specimen. This simulates deformation at depth within the Earth. An independent internal pore-fluid pressure also can be exerted. The rock specimen can be jacketed with a thin, impermeable sleeve (e.g., rubber or copper) to separate the external pressure medium from the internal pore fluids (if any). The specimen is typically a few centimetres in dimension.

Another apparatus for exerting high pressure on a sample was designed in 1968 by Akira Sawaoka, Naoto Kawai, and Robert Carmichael to give hydrostatic confining pressures up to 12 kilobars (1.2 gigapascal), additional directed stress, and temperatures up to a few hundred degrees Celsius. The specimen is positioned on the baseplate; the pressure is applied by driving in pistons with a hydraulic press. The end caps can be locked down to hold the pressure for time experiments and to make the device portable.

Apparatuses have been developed, typically using multianvil designs, which extend the range of static experimental conditions—at least for small specimens and limited times—to pressures as high as about 1,700 kilobars and temperatures of about 2,000 °C. Such work has been pioneered by researchers such as Peter M. Bell and Ho-Kwang Mao, who conducted studies at the Geophysical Laboratory of the Carnegie Institution in Washington, D.C. Using dynamic techniques (i.e., shock from explosive impact generated by gun-type designs), even higher pressures up to 7,000 kilobars (700 gigapascal)—which is nearly twice the pressure at the centre of the Earth and seven million times greater than the atmospheric pressure at the Earth's surface—can be produced for very short times. A leading figure in such ultrapressure work is A. Sawaoka at the Tokyo Institute of Technology.

In the upper crust of the Earth, hydrostatic pressure increases at the rate of about 320 bars per kilometre, and temperature increases at a typical rate of 20°–40 °C per kilometre, depending on recent crustal geologic history. Additional directed stress, as can be generated by large-scale crustal deformation (tectonism), can range up to 1 to 2 kilobars. This is approximately equal to the ultimate strength (before fracture) of solid crystalline rock at surface temperature and pressure. The stress released in a single major earthquake—a shift on a fault plane—is about 50–150 bars.

In studying the deformation of rocks one can start with the assumption of ideal behaviour: elastic strain and homogeneous and isotropic stress and strain. In reality, on a microscopic scale there are grains and pores in sediments and a fabric of crystals in igneous and metamorphic rocks. On a large scale, rock bodies exhibit physical and chemical variations and structural features. Furthermore, conditions such as extended length of time, confining pressure, and subsurface fluids affect the rates of change of deformation. Figure shows the generalized transition from brittle fracture through faulting to plastic-flow deformation in response to applied compressional stress and the progressive increase of confining pressure.

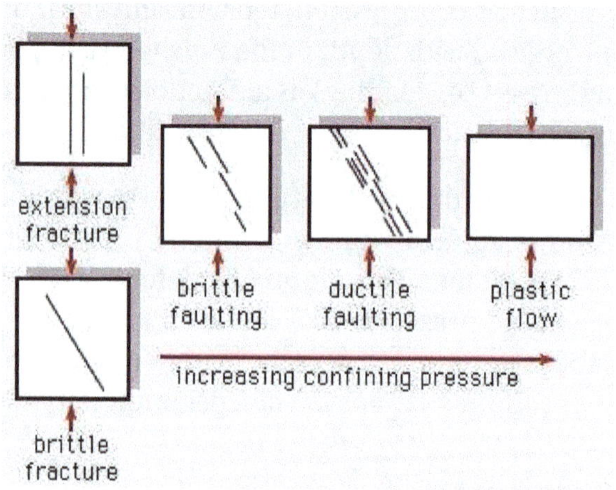

Deformation as affected by increased confining pressure.

Stress-strain Relationships

The deformation of materials is characterized by stress-strain relations. For elastic-behaviour materials, the strain is proportional to the load (i.e., the applied stress). The strain is immediate with stress and is reversible (recoverable) up to the yield point stress, beyond which permanent strain results. For viscous material, there is laminar (slow, smooth, parallel) flow; one must exert a force to maintain motion because of internal frictional resistance to flow, called the viscosity. Viscosity varies with the applied stress, strain rate, and temperature. In plastic behaviour, the material strains continuously (but still has strength) after the yield point stress is reached; however, beyond this point there is some permanent deformation. In elasticoviscous deformation, there is combined elastic and viscous behaviour. The material yields continuously (viscously) for a constant applied load. An example of such behaviour is creep, a slow, permanent, and continuous deformation occurring under constant load over a long time in such materials as crystals, ice, soil and sediment, and rocks at depth. In firmoviscous behaviour, the material is essentially solid but the strain is not immediate with application of stress; rather, it is taken up and released exponentially. A plasticoviscous material exhibits elastic behaviour for initial stress (as in plastic behaviour), but after the yield point stress is reached, it flows like a viscous fluid.

Some representative values of elastic constants and properties are listed in table. The coefficient of viscosity (η) is the ratio of applied stress to the rate of straining (change of strain with time). It is measured in units of poise; one poise equals one dyne-second per square centimetre.

Some typical values of elastic constants and properties		
	Elastic constants (at room temperature and pressure)	
Material	Young's modulus (in 10^6 bars)	shear modulus (in 10^6 bars)
Ice	0.1	0.03
Shale	0.2–0.3	0.15
Limestone	0.4–0.7	0.22–0.26
Granite	0.3–0.6	0.2
Basalt	0.7–0.9	0.3
Steel	2.1	0.83
Material	temperature (degrees Celsius)	coefficient of viscosity (poises)
Lava (Mount Vesuvius)	1,100 1,400	28,300 250
Lava (Oshima, Japan)	1,038 1,125	230,000 5,600
Andesite lava	1,400	150–1,500
Material	Compressive strength (at room temperature and pressure, in kilobars)	
Shale	0.8–1.8	
Sandstone	0.5–2	
Limestone	1–2	
Granite	1.7–2.5	
Basalt	1–3.4	

Rheology is the study of the flow deformation of materials. The concept of rheidity refers to the capacity of a material to flow, arbitrarily defined as the time required with a shear stress applied for the viscous strain to be 1,000 times greater than the elastic strain. It is thus a measure of the threshold of fluidlike behaviour. Although such behaviour depends on temperature, relative comparisons can be made. Some representative values of rheidity times are given in the table.

Rheidity threshold of fluidlike deformation	
Material	Approximate time
Ice (e.g., glacier)	2 weeks
Gypsum	1 year
Rock salt (e.g., saltdome)	10–20 years
Serpentine (a mafic silicate mineral)	10,000 years

Typical stress-strain (deformation) curves for rock materials are shown in Figure 8. The stress σ, compression in the figure, is force per unit area. The strain ε is fractional shortening of the specimen parallel to the applied compression; it is given here in percent. The brittle material behaves elastically nearly until the point of fracture (denoted X), whereas the ductile (plastically deformable) material is elastic up to the yield point but then has a range of plastic deformation before fracturing. The ability to undergo large permanent deformation before fracture is called ductility. For plastic deformation, the flow mechanisms are intracrystalline (slip and twinning within crystal grains), intercrystalline motion by crushing and fracture (cataclasis), and recrystallization by solutioning or solid diffusion.

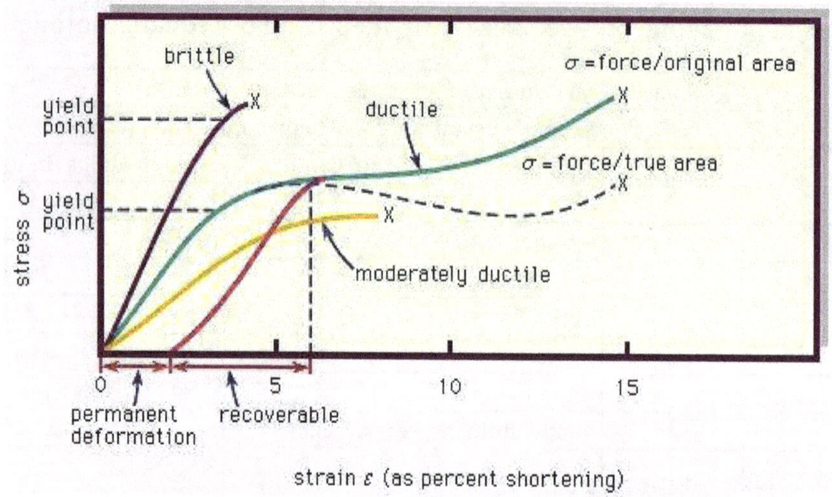

Typical stress-strain curves for rock materials. Each X represents the point of fracture for the corresponding material.

If the applied stress is removed while a ductile material is in the plastic range, part of the strain is recoverable (elastically), but there is permanent deformation. The ultimate strength is the highest point (stress) on a stress-strain curve, often occurring at fracture (which is the complete loss of cohesion). The strength of a material is its resistance to failure (destruction of structure) by flow or fracture; it is a measure of the stress required to deform a body. Typical compressive strengths (the stress required to cause failure under compression) are given in the Table.

Effect of Environmental Conditions

The behaviour and mechanical properties of rocks depend on a number of environmental conditions. (1) Confining pressure increases the elasticity, strength (e.g., yield point and ultimate fracture stress), and ductility. (2) Internal pore-fluid pressure reduces the effective stress acting on the sample, thus reducing the strength and ductility. The effective, or net, confining pressure is the external hydrostatic pressure minus the internal pore-fluid pressure. (3) Temperature lowers the strength, enhances ductility, and may enhance recrystallization. (4) Fluid solutions can enhance deformation, creep, and recrystallization. (5) Time is an influential factor as well. (6) The rate of loading (i.e., the rate at which stress is applied) influences mechanical properties. (7) Compaction, as would occur with burial to depth, reduces the volume of pore space for sedimentary rocks and the crack porosity for crystalline rocks.

Rocks, which are typically brittle at the Earth's surface, can undergo ductile deformation when buried and subjected to increased confining pressure and temperature for long periods of time. If stress exceeds their strength or if they are not sufficiently ductile, they will fail by fracture—as a crystal, within a bed or rock, on an earthquake fault zone, and so on—whereas with ductility they can flow and fold.

Some strengths for various rock types under different temperatures and confining pressures are listed in the Table. The plastic yield strength here is the stress at a 2 percent strain; the ultimate strength, as stated above, is the highest point (stress) on the stress-strain curve.

Rock strengths, with varying temperature and pressure				
Rock type	Temperature (°C)	Confining pressure (kilobars)	Plastic yield strength (kilobars)	Ultimate strength (kilobars)
Granite	500	5	10	11.5
	800	5	5	6
Gabbro	500	5	4	8
Peridotite	500	5	8	9
	800	5	5.5	8
Basalt	500	5	8	10
	800	5	2	2.5
Marble	24	2	2.5	5.5
	500	3	1	2
Limestone	24	2	1 5	5.5
	500	3	2.5	3
Dolomite	24	2	6	7
	500	5	4	6.5
Shale	24	2	1.5	2.5
Rock salt	24	1	0.5	1

An increase in confining pressure causes brittle fracture to become shear slippage and eventually causes flow (ductile) behaviour. This transition is also aided by higher temperature, decreased internal pore-fluid pressure, and slower strain rate.

Thermal Properties

Heat flow (or flux), q, in the Earth's crust or in rock as a building material, is the product of the temperature gradient (change in temperature per unit distance) and the material's thermal conductivity (k, the heat flow across a surface per unit area per unit time when a temperature difference exists in unit length perpendicular to the surface). Thus,

$$q = \frac{dT}{dz} \times k.$$

The units of the terms in this equation are given below, expressed first in the centimetre-gram-second (cgs) system and then in the International System of Units (SI) system, with the conversion factor from the first to the second given between them.

q, heat flow:

calories per square

centimetre per second $\times \frac{1}{23.9} \times 10^{-6}$ = watts per square metre $(cal / cm^2.\sec)$ W/m^2

$\frac{dT}{dz}$, temperature gadient:

degrees Celsiusper kilometre($^\circ C / km$, practical unit for Earth)	$\times 10^{-3}$	degrees Celsius per metre ($^\circ C / m$)

k, thermal conductivity:

calories per square centimetre per scond perdegree celsius $(cal / cm^2.\sec.^\circ C)$	$\times \frac{1}{23.9} \times 10^{-3}$ =	Watts per metre per degree celsius $(W/m.^\circ C)$

Thermal Conductivity

Thermal conductivity can be determined in the laboratory or in situ, as in a borehole or deep well, by turning on a heating element and measuring the rise in temperature with time. It depends on several factors: (1) chemical composition of the rock (i.e., mineral content), (2) fluid content (type and degree of saturation of the pore space); the presence of water increases the thermal conductivity (i.e., enhances the flow of heat), (3) pressure (a high pressure increases the thermal conductivity by closing cracks which inhibit heat flow), (4) temperature, and (5) isotropy and homogeneity of the rock.

For crystalline silicate rocks—the dominant rocks of the "basement" crustal rocks—the lower values are typical of ones rich in magnesium and iron (e.g., basalt and gabbro) and the higher values are typical of those rich in silica (quartz) and alumina (e.g., granite). These values result because the thermal conductivity of quartz is relatively high, while that for feldspars is low.

Thermal Expansion

The change in dimension—linear or volumetric—of a rock specimen with temperature is expressed in terms of a coefficient of thermal expansion. This is given as the ratio of dimension change (e.g., change in volume) to the original dimension (volume, V) per unit of temperature (T) change:

$$\frac{1}{V}\frac{\Delta V}{\Delta T}$$

Thermal expansion of rocks	
Rock type	Linear expansion coefficient ($\times 10^{-6}$ per degree Celsius)
Granite and rhyolite	8 ± 3
Andesite and diorite	7 ± 2
Basalt, gabbro, and diabase	5.4 ± 1
Sandstone	10 ± 2
Limestone	8 ± 4
Marble	7 ± 2
Slate	9 ± 1

$$\frac{1}{L}\frac{\Delta L}{\Delta T},$$

Where, L represents length. All data are based on at least three samples.

Radioactive Heat Generation

The spontaneous decay (partial disintegration) of the nuclei of radioactive elements provides decay particles and energy. The energy, composed of emission kinetic energy and radiation, is converted to heat; it has been an important factor in affecting the temperature gradient and thermal evolution of the Earth. Deep-seated elevated temperatures provide the heat that causes rock to deform plastically and to move, thus generating to a large extent the processes of plate tectonics—plate motions, seafloor spreading, continental drift, and subduction—and most earthquakes and volcanism.

Some elements, or their isotopes (nuclear species with the same atomic number but different mass numbers), decay with time. These include elements with an atomic number greater than 83—of which the most important are uranium-235, uranium-238, and thorium-232—and a few with a lower atomic number, such as potassium-40.

The heat generated within rocks depends on the types and abundances of the radioactive elements and their host minerals. Such heat production, A, is given in calories per cubic centimetre per second, or 1 calorie per gram per year = 4.186 × 107 ergs per gram per year = 1.327 ergs per gram per second. The rate of radioactive decay, statistically an exponential process, is given by the half-life, t1/2. The half-life is the time required for half the original radioactive atoms to decay for a particular isotope.

Some radioactive decay series are listed in the Table. The isotopic abundance is the percent of the natural element that exists as that particular radioactive isotope; for example, 99.28 percent of natural uranium is U-238, and 100 percent of thorium is the radioactive Th-232. The final product is the end result of the process (usually multistage) of disintegration. For the rocks, the typical content is given for uranium and thorium (in parts per million [ppm] of weight) and for potassium (in weight percent). The heat production of natural uranium is close to that for the isotope U-238, since almost all natural uranium is of that isotopic species.

The radioactive elements are more concentrated in the continental upper-crust rocks that are rich in quartz (i.e., felsic, or less mafic). This results because these rocks are differentiated by partial melting of the upper-mantle and oceanic-crust rock. The radioactive elements tend to be preferentially driven off from these rocks for geochemical reasons. A compilation of heat productivities of various rock types is given in the table.

Electrical Properties

The electrical nature of a material is characterized by its conductivity (or, inversely, its resistivity) and its dielectric constant, and coefficients that indicate the rates of change of these with temperature, frequency at which measurement is made, and so on. For rocks with a range of chemical composition as well as variable physical properties of porosity and fluid content, the values of electrical properties can vary widely.

Resistance (R) is defined as being one ohm when a potential difference (voltage; V) across a specimen of one volt magnitude produces a current (i) of one ampere; that is, V = Ri. The electrical resistivity (ρ) is an intrinsic property of the material. In other words, it is inherent and not dependent on sample size or current path. It is related to resistance by R = ρL/A where L is the length of specimen, A is the cross-sectional area of specimen, and units of ρ are ohm-centimetre; 1 ohm-centimetre equals 0.01 ohm-metre. The conductivity (σ) is equal to $1/\rho$ $\text{ohm}^{-1}\cdot\text{centimetre}^{-1}$ (or termed mhos/cm). In SI units, it is given in mhos/metre, or siemens/metre.

Some representative values of electrical resistivity for rocks and other materials are listed in the Table. Materials that are generally considered as "good" conductors have a resistivity of $10^{-5}-10$ ohm – centimetre ($10^{-7}-10^{-1}$ ohm – metre) and a conductivity of $10-10^{7}$ mhos/metre. Those that are classified as intermediate conductors have a resistivity of $100-10^{9}$ ohm-centimetre ($1-10^{7}$ ohm-metre) and a conductivity of $10^{-7}-1$ mhos/metre. "Poor" conductors, also known as insulators, have a resistivity of $10^{10}-10^{17}$ ohm-centimetre ($10^{8}-10^{15}$ ohm-metre) and a conductivity of $0^{-15}-10^{-8}$. Seawater is a much better conductor (i.e., it has lower resistivity) than fresh water owing to its higher content of dissolved salts; dry rock is very resistive. In the subsurface, pores are typically filled to some degree by fluids. The resistivity of materials has a wide range—copper is, for example, different from quartz by 22 orders of magnitude.

Typical resistivities		
Material		Resistivity (ohm-centimetre)
Seawater (18 °c)		21
Uncontaminated surface water		$2(10^{4})$
Distilled water		$0.2–1(10^{6})$
Water (4 °c)		$9(10^{6})$
Ice		$3(10^{8})$
Rocks in situ		
Sedimentary	Clay, soft shale	$100–5(10^{3})$
	Hard shale	$7–50(10^{3})$
	Sand	$5–40(10^{3})$
	Sandstone	$(10^{4})–(10^{5})$
	Glacial moraine	$1–500(10^{3})$
	Porous limestone	$1–30(10^{4})$
	Dense limestone	$>(10^{6})$
	Rock salt	$(10^{8})–(10^{9})$
Igneous		$5(10^{4})–(10^{8})$
Metamorphic		$5(10^{4})–5(10^{9})$

Rocks in laboratory	
Dry granite	10^{12}
Minerals	
Copper (18 °c)	$1.7(10^{-6})$
Graphite	$5–500(10^{-4})$
Pyrrhotite	0.1–0.6
Magnetite crystals	0.6–0.8
Pyrite ore	$1–(10^{5})$
Magnetite ore	$(10^{2})–5(10^{5})$
Chromite ore	$>10^{6}$
Quartz (18 °c)	$(10^{14})–(10^{16})$

For high-frequency alternating currents, the electrical response of a rock is governed in part by the dielectric constant, ε. This is the capacity of the rock to store electric charge; it is a measure of polarizability in an electric field. In cgs units, the dielectric constant is 1.0 in a vacuum. In SI units, it is given in farads per metre or in terms of the ratio of specific capacity of the material to specific capacity of vacuum (which is 8.85×10^{-12} farads per metre). The dielectric constant is a function of temperature, and of frequency, for those frequencies well above 100 hertz (cycles per second).

Electrical conduction occurs in rocks by (1) fluid conduction—i.e., electrolytic conduction by ionic transfer in briny pore water—and (2) metallic and semiconductor (e.g., some sulfide ores) electron conduction. If the rock has any porosity and contained fluid, the fluid typically dominates the conductivity response. The rock conductivity depends on the conductivity of the fluid (and its chemical composition), degree of fluid saturation, porosity and permeability, and temperature. If rocks lose water, as with compaction of clastic sedimentary rocks at depth, their resistivity typically increases.

Magnetic Properties

The magnetic properties of rocks arise from the magnetic properties of the constituent mineral grains and crystals. Typically, only a small fraction of the rock consists of magnetic minerals. It is this small portion of grains that determines the magnetic properties and magnetization of the rock as a whole, with two results: (1) the magnetic properties of a given rock may vary widely within a given rock body or structure, depending on chemical inhomogeneities, depositional or crystallization conditions, and what happens to the rock after formation; and (2) rocks that share the same lithology (type and name) need not necessarily share the same magnetic characteristics. Lithologic classifications are usually based on the abundance of dominant silicate minerals, but the magnetization is determined by the minor fraction of such magnetic mineral grains as iron oxides. The major rock-forming magnetic minerals are iron oxides and sulfides.

Although the magnetic properties of rocks sharing the same classification may vary from rock to rock, general magnetic properties do nonetheless usually depend on rock type and overall composition. The magnetic properties of a particular rock can be quite well understood provided one has specific information about the magnetic properties of crystalline materials and minerals, as well as about how those properties are affected by such factors as temperature, pressure, chemical composition, and the size of the grains. Understanding is further enhanced by information about how the properties of typical rocks are dependent on the geologic environment and how they vary with different conditions.

Applications of the Study of Rock Magnetization

An understanding of rock magnetization is important in at least three different areas: prospecting, geology, and materials science. In magnetic prospecting, one is interested in mapping the depth, size, type, and inferred composition of buried rocks. The prospecting, which may be done from ground surface, ship, or aircraft, provides an important first step in exploring buried geologic structures and may, for example, help identify favourable locations for oil, natural gas, and economic mineral deposits.

Rock magnetization has traditionally played an important role in geology. Paleomagnetic work seeks to determine the remanent magnetization and thereby ascertain the character of the Earth's field when certain rocks were formed. The results of such research have important ramifications in stratigraphic correlation, age dating, and reconstructing past movements of the Earth's crust. Indeed, magnetic surveys of the oceanic crust provided for the first time the quantitative evidence needed to cogently demonstrate that segments of the crust had undergone large-scale lateral displacements over geologic time, thereby corroborating the concepts of continental drift and seafloor spreading, both of which are fundamental to the theory of plate tectonics.

The understanding of magnetization is increasingly important in materials science as well. The design and manufacture of efficient memory cores, magnetic tapes, and permanent magnets increasingly rely on the ability to create materials having desired magnetic properties.

Basic Types of Magnetization

There are six basic types of magnetization: (1) diamagnetism, (2) paramagnetism, (3) ferromagnetism, (4) antiferromagnetism, (5) ferrimagnetism, and (6) superparamagnetism.

Diamagnetism arises from the orbiting electrons surrounding each atomic nucleus. When an external magnetic field is applied, the orbits are shifted in such a way that the atoms set up their own magnetic field in opposition to the applied field. In other words, the induced diamagnetic field opposes the external field. Diamagnetism is present in all materials, is weak, and exists only in the presence of an applied field. The propensity of a substance for being magnetized in an external field is called its susceptibility (k) and it is defined as J/H, where J is the magnetization (intensity) per unit volume and H is the strength of the applied field. Since the induced field always opposes the applied field, the sign of diamagnetic susceptibility is negative. The susceptibility of a diamagnetic substance is on the order of -10^{-6} electromagnetic units per cubic centimetre (emu/cm^3). It is sometimes denoted κ for susceptibility per unit mass of material.

Paramagnetism results from the electron spin of unpaired electrons. An electron has a magnetic dipole moment—which is to say that it behaves like a tiny bar magnet—and so when a group of electrons is placed in a magnetic field, the dipole moments tend to line up with the field. The effect augments the net magnetization in the direction of the applied field. Like diamagnetism, paramagnetism is weak and exists only in the presence of an applied field, but since the effect enhances the applied field, the sign of the paramagnetic susceptibility is always positive. The susceptibility of a paramagnetic substance is on the order of 10^{-4} to 10^{-6} emu/cm^3.

Ferromagnetism also exists because of the magnetic properties of the electron. Unlike paramagnetism, however, ferromagnetism can occur even if no external field is applied. The magnetic dipole moments of the atoms spontaneously line up with one another because it is energetically favourable for them to do so. A remanent magnetization can be retained. Complete alignment of the dipole moments would take place only at a temperature of absolute zero (0 kelvin [K], or -273.15 °C). Above absolute zero, thermal motions begin to disorder the magnetic moments. At a temperature called the Curie temperature, which varies from material to material, the thermally induced disorder overcomes the alignment, and the ferromagnetic properties of the substance disappear. The susceptibility of ferromagnetic materials is large and positive. It is on the order of 10 to 10^4 emu/cm^3. Only a few materials—iron, cobalt, and nickel—are ferromagnetic in the strict sense of the word and have a strong residual magnetization. In general usage, particularly in engineering, the term ferromagnetic is frequently applied to any material that is appreciably magnetic.

Antiferromagnetism occurs when the dipole moments of the atoms in a material assume an antiparallel arrangement in the absence of an applied field. The result is that the sample has no net magnetization. The strength of the susceptibility is comparable to that of paramagnetic materials. Above a temperature called the Néel temperature, thermal motions destroy the antiparallel arrangement, and the material then becomes paramagnetic. Spin-canted (anti)ferromagnetism is a special condition which occurs when antiparallel magnetic moments are deflected from the antiferromagnetic plane, resulting in a weak net magnetism. Hematite (α-Fe_2O_3) is such a material.

Ferrimagnetism is an antiparallel alignment of atomic dipole moments which does yield an appreciable net magnetization resulting from unequal moments of the magnetic sublattices. Remanent magnetization is detectable. Above the Curie temperature the substance becomes paramagnetic. Magnetite (Fe_3O_4), which is the most magnetic common mineral, is a ferrimagnetic substance.

Superparamagnetism occurs in materials having grains so small (about 100 angstroms) that any cooperative alignment of dipole moments is overcome by thermal energy.

Types of Remanent Magnetization

Rocks and minerals may retain magnetization after the removal of an externally applied field, thereby becoming permanent weak magnets. This property is known as remanent magnetization and is manifested in different forms, depending on the magnetic properties of the rocks and minerals and their geologic origin and history. Delineated below are the kinds of remanent magnetization frequently observed.

CRM (chemical, or crystallization, remanent magnetization) can be induced after a crystal is formed and undergoes one of a number of physicochemical changes, such as oxidation or reduction, a phase change, dehydration, recrystallization, or precipitation of natural cements. The induction, which is particularly important in some (red) sediments and metamorphic rocks, typically takes place at constant temperature in the Earth's magnetic field.

DRM (depositional, or detrital, remanent magnetization) is formed in clastic sediments when fine particles are deposited on the floor of a body of water. Marine sediments, lake sediments, and some clays can acquire DRM. The Earth's magnetic field aligns the grains, yielding a preferred direction of magnetization.

IRM (isothermal remanent magnetization) results from the application of a magnetic field at a constant (isothermal) temperature, often room temperature.

NRM (natural remanent magnetization) is the magnetization detected in a geologic in situ condition. The NRM of a substance may, of course, be a combination of any of the other remanent magnetizations.

PRM (pressure remanent, or piezoremanent, magnetization) arises when a material undergoes mechanical deformation while in a magnetic field. The process of deformation may result from hydrostatic pressure, shock impact (as produced by a meteorite striking the Earth's surface), or directed tectonic stress. There are magnetization changes with stress in the elastic range, but the most pronounced effects occur with plastic deformation when the structure of the magnetic minerals is irreversibly changed.

TRM (thermoremanent magnetization) occurs when a substance is cooled, in the presence of a magnetic field, from above its Curie temperature to below that temperature. This form of magnetization is generally the most important, because it is stable and widespread, occurring in igneous and sedimentary rocks. TRM also can occur when dealing exclusively with temperatures below the Curie temperature. In PTRM (partial thermoremanent magnetization) a sample is cooled from a temperature below the Curie point to yet a lower temperature.

VRM (viscous remanent magnetization) results from thermal agitation. It is acquired slowly over time at low temperatures and in the Earth's magnetic field. The effect is weak and unstable but is present in most rocks.

Hysteresis and Magnetic Susceptibility

The concept of hysteresis is fundamental when describing and comparing the magnetic properties of rocks. Hysteresis is the variation of magnetization with applied field and illustrates the ability of a material to retain its magnetization, even after an applied field is removed. Figure illustrates this phenomenon in the form of a plot of magnetization (J) versus applied field (H_{ex}). J_s is the saturation (or "spontaneous") magnetization when all the magnetic moments are aligned in their configuration of maximum order. It is temperature-dependent, reaching zero at the Curie temperature. $J_{r,sat}$ is the remanent magnetization that remains when a saturating (large) applied field is removed, and J_r is the residual magnetization left by some process apart from IRM saturation, as, for example, TRM. H_c is the coercive field (or force) that is required to reduce $J_{r,sat}$ to zero, and $H_{c,r}$ is the field required to reduce J_r to zero.

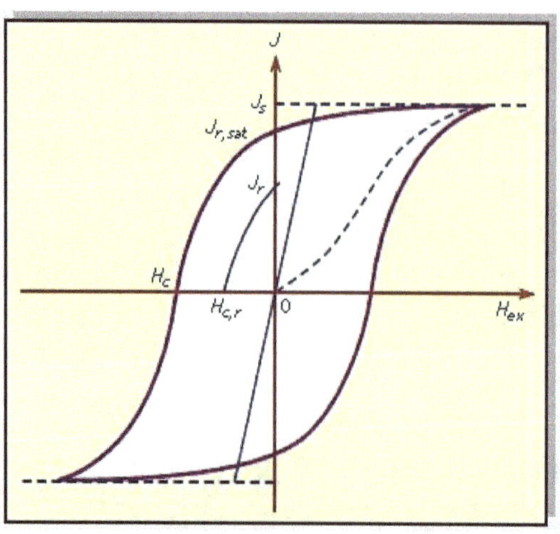

General magnetic hysteresis curve, showing magnetization (J) as a function of the external field (H_{ex}). J_s is the saturation (or "spontaneous") magnetization; $J_{r,sat}$ is the remanent magnetization that remains after a saturating applied field is removed; J_r is the residual magnetization left by some magnetization process other than IRM saturation; H_c is the coercive field; and $H_{c,r}$ is the field necessary to reduce J_r to zero.

Magnetic susceptibility is a parameter of considerable diagnostic and interpretational use in the study of rocks. This is true whether an investigation is being conducted in the laboratory or magnetic fields over a terrain are being studied to deduce the structure and lithologic character of buried rock bodies. Susceptibility for a rock type can vary widely, depending on magnetic mineralogy, grain size and shape, and the relative magnitude of remanent magnetization present, in addition to the induced magnetization from the Earth's weak field. The latter is given as $J_{induced} = kH_{ex}$, where k is the (true) magnetic susceptibility and H_{ex} is the external (i.e., the Earth's) magnetic field. If there is an additional remanent magnetization with its ratio (Q_n) to induced magnetization being given by,

$$Q_n = \frac{J_{remanent}}{kH_{ex}}$$

Then the total magnetization is,

$$\begin{aligned} J &= J_{induced} + J_{remanent} \\ &= kH_{ex} + Q_n kh_{ex} \\ &= k_{app} H_{ex}, \end{aligned}$$

Where, k_{app}, the "apparent" magnetic susceptibility, is $k(1+Q_n)$.

Magnetic Minerals and Magnetic Properties of Rocks

The major rock-forming magnetic minerals are the following iron oxides: the titanomagnetite series, $xFe_2TiO_4 \cdot (1-x)Fe_3O_4$, where Fe_3O_4 is magnetite, the most magnetic mineral; the ilmenohematite series, $yFeTiO_3 \cdot (1-y)Fe_2O_3$, where $\alpha - Fe_2O_3$ (in its rhombohedral structure) is hematite; maghemite, $\gamma - Fe_2O_3$ (in which some iron atoms are missing in the hematite structure); and limonite (hydrous iron oxides). They also include sulfides—namely, the pyrrhotite series, $yFeS \cdot (1-y)Fe_{1-x}S$.

The table gives some typical values of the apparent susceptibility for various rock types, which usually include some remanent as well as induced magnetization. Values are higher for mafic igneous rocks, especially as the content of magnetite increases.

Approximate "apparent susceptibilities" for rock types.	
Rock	Apparent magnetic susceptibility (electromagnetic units per cubic centimetre)
Iron ores	Over 0.1
Basalt	10^{-2}

Approximate "apparent susceptibilities" for rock types.	
Rock	Apparent magnetic susceptibility (electromagnetic units per cubic centimetre)
Andesite	10^{-3}
Dacite	10^{-4}
Metamorphic rocks	10^{-4}
Sedimentary rocks	10^{-5}

A distribution of measured (true) susceptibilities for various rock types is shown in the table. Basic refers to those rocks high in iron and magnesium silicates and magnetite, extrusive means formed by cooling after extruding onto the land surface or seafloor. The data in each category are based on at least 45 samples.

Measured susceptibilities for rock types				
Rock type	% Of samples with magnetic susceptibility (in 10–6 electromagnetic unit per cubic centimetre)			
	Less than 100	100–1,000	1,000–4,000	Greater than 4,000
Basic extrusive (e.g., Basalt)	5	29	47	19
Basic intrusive (e.g., Gabbro)	24	27	28	21
Granite	60	23	16	1
Metamorphic (gneiss, schist, slate)	71	22	7	0
Sedimentary	73	19	4	4

Igneous Rock

Igneous rocks form from the cooling of magma – molten materials in the earth's crust. The terminology Igneous means fire or heat. In this sense, igneous rocks are formed when molten rock (magma) solidifies either underneath the earth crust to form plutonic (intrusive) igneous rocks or on the surface of the earth to form volcanic (extrusive) igneous rocks.

They are simply the rocks formed through heating then followed by cooling. The heated material is the molten rock which is made up of partial or complete melting of previously existent rocks in the earth's crust that are consistently subjected to intense heat, high pressure changes, and alterations in composition.

"Igneous rock is one of the three main rock types, the others being sedimentary and metamorphic. Igneous rock is formed through the cooling and solidification of magma or lava. Igneous rock may form with or without crystallization, either below the surface as intrusive (plutonic) rocks or on the surface as extrusive (volcanic) rocks."

Igneous-rock-the-earth-vomited-this-rock.

Formation of Igneous Rocks

Molten materials are found below the earth crust and are normally subjected to extreme pressure and temperatures – up to 1200° Celsius. Because of the high temperatures and pressure changes, the molten materials sometimes shoot up to the surface in the form of volcanic eruption and they cool down to form volcanic or extrusive igneous rocks.

Alternatively, some of the molten materials may cool underneath the earth surface very slowly to form plutonic or intrusive igneous rocks. It is because of the extreme heat levels and changes in pressure that igneous rocks do not contain organic matter or fossils. The molten minerals interlock and crystallize as the melt cools and form solid materials.

In the long-run, the melt forms a cool hard rock made up of crystals with no open spaces and don't exhibit any desirable grain alignment. The rocks may be made up entirely of one mineral or various minerals, and their sizes are determined by the cooling process. Rapid cooling results in smaller crystals while slow cooling results in large crystals.

Types of Igneous Rocks

Igneous rocks are of two types, intrusive (plutonic rocks) and extrusive (volcanic rocks).

Intrusive Igneous Rocks

Intrusive igneous rocks are formed when the magma cools off slowly under the earth's crust and hardens into rocks. Gabbro and granite are examples of intrusive igneous rocks. Intrusive rocks are very hard in nature and are often coarse-grained.

Extrusive Igneous Rocks

Extrusive igneous rocks are formed when molten magma spill over to the surface as a result of volcanic eruption. The magma on the surface (lava) cools faster on the surface to form igneous rocks that are fine grained. Examples of such kind of rocks include pumice, basalt, or obsidian.

Examples of Igneous Rocks

There are more than 700 known types of igneous rocks and most of them are formed under the earth's crust since volcanic events are not all that frequent. On this basis, we are going to look at the commonly identified types of igneous rocks, both intrusive and extrusive. They include:

Granite

Granites are the light-colored and coarse-grained igneous rocks. They are intrusive rocks and they contain three major minerals including feldspar, mica and quartz. They appear pinkish, gray or tan depending on the grain sizes and concentrations and grain sizes of the three minerals. Granites are often used in construction activities because of its strength and presence in great quantities.

Gabbro

Gabbros are the dark-colored and coarse-grained igneous rocks. They are intrusive rocks and they are made up of mineral elements including pyroxene, feldspar, and sometimes olivine. They appear grey in color with minute dark spots all over them.

Basalt

Basalts are the dense dark-grey colored and fine-grained igneous rocks. They are extrusive and are chiefly composed of plagioclase and pyroxene. They are the commonest type of solidified lava. Basalts are also frequently used in building and construction.

Diorite

Diorites are the coarse-grained igneous rocks just like the Gabbros and Granite. They are intrusive and contain a mixture of minerals including hornblende, pyroxene, feldspar and sometimes quartz. They appear light-colored with some dark spots.

Andesite

Andesites are light grey-colored and fine-grained igneous rocks. They are extrusive rocks which are mainly made up of plagioclase minerals mixed up with biotite, pyroxene, and hornblende.

Obsidian

Obsidians are the dense dark-colored and smooth igneous rocks. They are the granitic or volcanic glass formed by the rapid cooling of lava without crystallization. Obsidians usually appear dark, but transparent in thin pieces.

Pegmatite

Pegmatite is a form of igneous rock consisting of extremely coarse granite resulting from the crystallization of magma rich in rare elements. Pegmatite is light-colored and often contains rare minerals that are not present in other parts of the magma chamber.

Pumice

Pumice is a light-colored vesicular and porous like igneous rock that forms as a result of very fast solidification of molten rock material. The vesicular and porous like texture is due to the gas trapped within the melt during the rapid solidification. Pumice stones are commonly used as abrasive materials.

Peridotite

Peridotite is a dark coarse-grained igneous rock consisting principally of olivine. It is an intrusive rock and may also contain some small amount of minerals such as feldspar, quartz, amphibole, pyroxene, or quartz.

Rhyolite

Rhyolites are light-colored and fine-grained igneous rocks. They are extrusive igneous rocks and they are typically composed feldspar and quartz minerals. They often have a smooth surface.

Welded Tuff

Welded Tuffs are extrusive igneous rocks made up of the materials discharged during a volcanic event. The discharged materials fall on the earth surface after which they lithify into rocks. Welded Tuffs are mostly composed of volcanic ash and at times include large size pieces such as clinkers.

Fire Opal

Fire opals are the igneous rocks that are flaming orange or yellow and red in color. At times they are found in rhyolite and are categorized as one of the most spectacular igneous geological rocks. They are formed long after the cooling of rhyolite whereby silica-rich ground water moves into the rocks and sometimes deposit gems such as topaz, agate, opal, jasper, or red beryl in the rock's cavities.

Scoria

Scoria is a dark-colored vesicular and porous like igneous rock. It is an extrusive rock and the vesicles or porous like composition is due to the gas trapped within the melt during cooling and solidification. Scoria usually forms as a bubbling crust on top of the lava as it flows down from the volcanic vent and eventually solidifies with some air trapped inside.

Sedimentary Rock

Sedimentary rocks are the second great rock class. Whereas igneous rocks are born hot, sedimentary rocks are born cool at the Earth's surface, mostly under water. They usually consist of layers or strata; hence they are also called stratified rocks. Depending on what they're made of, sedimentary rocks fall into one of three types.

The main thing about sedimentary rocks is that they were once sediment — mud and sand and gravel and clay — and were not greatly changed as they turned into rock. The following traits are all related to that:

- They're generally arranged in layers of sandy or clayey material (strata) like those you'll see in excavations or a hole dug in a sand dune.

- They're usually the color of sediment, that is, light brown to light gray.
- They may preserve signs of life and surface activity, like fossils, tracks, ripple marks and so on.

Clastic Sedimentary Rocks

The most common set of sedimentary rocks consists of the granular materials that occur in sediment. Sediment mostly consists of surface minerals — quartz and clays — that are made by the physical breakdown and chemical alteration of rocks. These are carried away by water or the wind and laid down in a different place. Sediment may also include pieces of stones and shells and other objects, not just grains of pure minerals. Geologists use the word clasts to denote particles of all these kinds, and rocks made of clasts are called clastic rocks.

Look around you at where the world's clastic sediment goes: sand and mud are carried down rivers to the sea, mostly. Sand is made of quartz, and mud is made of clay minerals. As these sediments are steadily buried over geologic time, they get packed together under pressure and low heat, not much more than 100 °C. In these conditions the sediment is cemented into rock: sand becomes sandstone and clay become shale. If gravel or pebbles are part of the sediment, the rock that forms is conglomerate. If the rock is broken and recemented together, it is called breccia.

It's worth noting that some rocks commonly lumped in the igneous category are actually sedimentary. Tuff is consolidated ash that has fallen from the air in volcanic eruptions, making it just as sedimentary as a marine claystone. There is some movement in the profession to recognize this truth.

Organic Sedimentary Rocks

Another type of sediment actually arises in the sea as microscopic organisms — plankton — build shells out of dissolved calcium carbonate or silica. Dead plankton steadily shower their dust-sized shells onto the seafloor, where they accumulate in thick layers. That material turns to two more rock types, limestone (carbonate) and chert (silica). These are called organic sedimentary rocks, although they're not made of organic material as a chemist would define it.

Another type of sediment forms where dead plant material builds up into thick layers. With a small degree of compaction, this becomes peat; after much longer and deeper burial, it becomes coal. Coal and peat are organic in both the geological and the chemical sense.

Although peat is forming in parts of the world today, the great beds of coal that we mine formed during past ages in enormous swamps. There are no coal swamps around today because conditions do not favor them. The sea needs to be much higher. Most of the time, geologically speaking, the sea is hundreds of meters higher than today, and most of the continents are shallow seas. That's why we have sandstone, limestone, shale and coal over most of the central United States and elsewhere on the world's continents. (Sedimentary rocks also become exposed when the land rises. This is common around the edges of the Earth's lithospheric plates.

Chemical Sedimentary Rocks

These same ancient shallow seas sometimes allowed large areas to become isolated and begin drying up. In that setting, as the seawater grows more concentrated, minerals begin to come out of solution (precipitate), starting with calcite, then gypsum, then halite. The resulting rocks are certain limestones, gypsum rock, and rock salt respectively. These rocks, called the evaporite sequence, are also part of the sedimentary clan.

In some cases, chert can also form by precipitation. This usually happens below the sediment surface, where different fluids can circulate and interact chemically.

Diagenesis: Underground Changes

All kinds of sedimentary rocks are subject to further changes during their stay underground. Fluids may penetrate them and change their chemistry; low temperatures and moderate pressures may change some of the minerals into other minerals. These processes, which are gentle and do not deform the rocks, are called diagenesis as opposed to metamorphism (although there is no well-defined boundary between the two).

The most important types of diagenesis involve the formation of dolomite mineralization in limestones, the formation of petroleum and of higher grades of coal, and the formation of many types of ore bodies. The industrially important zeolite minerals also form by diagenetic processes.

Metamorphic Rock

Metamorphic rocks are formed through the transformation of pre-existing rocks in a process known as metamorphism. The original rock, or protolith, is subjected to heat and pressure which cause physical, chemical and mineralogical changes to the rock. Protoliths may be igneous, sedimentary or pre-existing metamorphic rocks.

Metamorphic rocks are formed within the Earth's crust. Changing temperature and pressure conditions may result in changes to the mineral assemblage of the protolith. Metamorphic rocks are eventually exposed at the surface by uplift and erosion of the overlying rock.

There are two main types of metamorphism: regional metamorphism and contact, or thermal, metamorphism. Metamorphic rocks are categorised by texture and mineralogy. Specific zones of temperature and pressure define different metamorphic facies. Rocks containing minerals diagnostic of each facies can be linked to the temperature and pressure that characterise that facies. For example, blueschist and eclogite facies rocks form under low to medium temperatures and high pressures. Such conditions as usually associated with subduction zones. Hornfels facies rocks typically occur under the medium to high temperature and low pressure conditions associated with contact aureoles.

Each metamorphic facies is associated with a group of index minerals from which it can be identified. The exact set of indicator minerals from each group that is present in a particular instance depends on the chemical composition of the protolith. Information about the mineralogy of each metamorphic facies may be found in Grotzinger and Jordan.

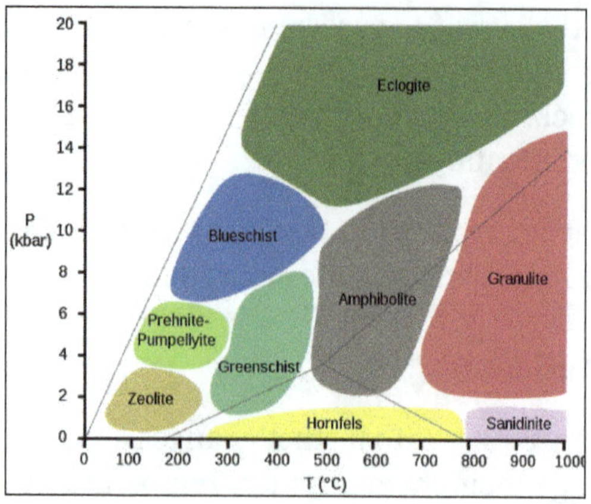

Metamorphic facies. Greenschist, amphibolite and granulite conditions are typically associated with regional metamorphism. Blueschist and eclogite are associated with the type of regional metamorphism that occurs at subduction zones. Hornfels and sanidinite are typically associated with contact metamorphism.

Regional Metamorphism

Regional metamorphism usually results in the formation of rocks that are strongly foliated, such as slates, schists, and gneisses. The differential stress needed to produce such foliated rocks usually results from tectonic forces that produce compressional stresses in the rocks such as occur when two continental masses collide. Thus, regionally metamorphosed rocks are found in the cores of fold/thrust mountain belts or in eroded mountain ranges. Compressive stresses result in folding of rock and thickening of the crust, which results in some of the rocks being pushed to deeper levels where they are subjected to higher temperatures and pressures.

The pressure and heat which drive metamorphism come from three sources: the internal heat of the Earth; the weight of the overlying rock; and the horizontal pressures developed as rocks become deformed.

Temperature within the crust increases with increasing depth from the surface by 20 to 60 °C per kilometre of depth; 30 °C per km is considered to be the average. For example, at a depth of 15km the temperature will be about 450 °C. The pressure at 15km depth is derived from the weight of the overlying rock, and is about 4000 times the pressure at the Earth's surface.

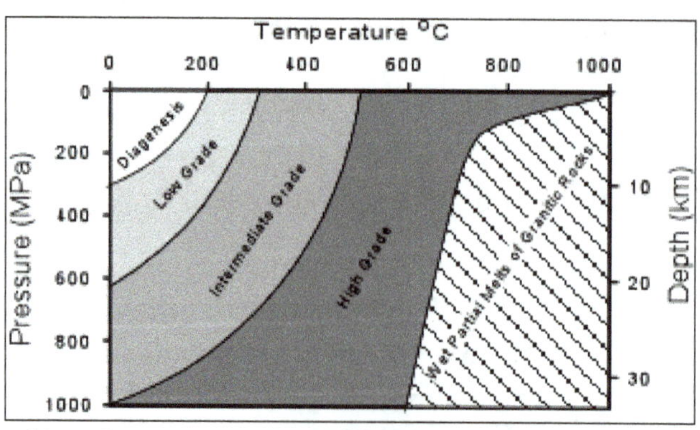

Grades of metamorphic rock. A metamorphic rock which has experienced a temperature of 600 °C and a pressure of 600MPa is termed a "high-grade metamorphic rock".

Metamorphic rocks formed in the shallow crust are referred to as low-grade metamorphic rocks; those at higher temperatures and pressures as high-grade metamorphic rocks. Through uplift and erosion, these rocks are eventually exposed at the Earth's surface.

Contact Metamorphism

Contact metamorphism typically occurs around intrusive igneous rocks, and is the result of temperature increase associated with the introduction of hot magma into the colder country rocks. The area surrounding an intrusion, where metamorphism is present, is called a metamorphic aureole. Metamorphism is greatest closest to the intrusion and dissipates as distance from the intrusion increases. The size of the aureole depends on the temperature of the igneous intrusion, the size of the intrusion and the temperature of the country rock. Dykes may have a small band of adjacent rock which has been locally metamorphosed, but the effects of this metamorphism are usually minimal. Large intrusions, such as granite batholiths, may have metamorphic aureoles covering a significant area. Where the country rocks are pelitic in composition (i.e. they are formed from small silt particles), the rock closest to the intrusion is "baked", forming a hard, splintery type of rock referred to as a hornfels. Hornfels may have a spotted appearance due to the growth of minerals such as cordierite and andalusite. In rocks of different compositions other minerals will form. For example olivine may form in an impure limestone.

The metamorphic grade of an aureole zone is quantified by which minerals occur within that zone. Shale closest to an intrusion may be altered to hornfels and contain minerals such as sillimanite or cordierite. Further from the intrusion, the same shale may seem reasonably unaltered apart from the development of some crystals of andalusite.

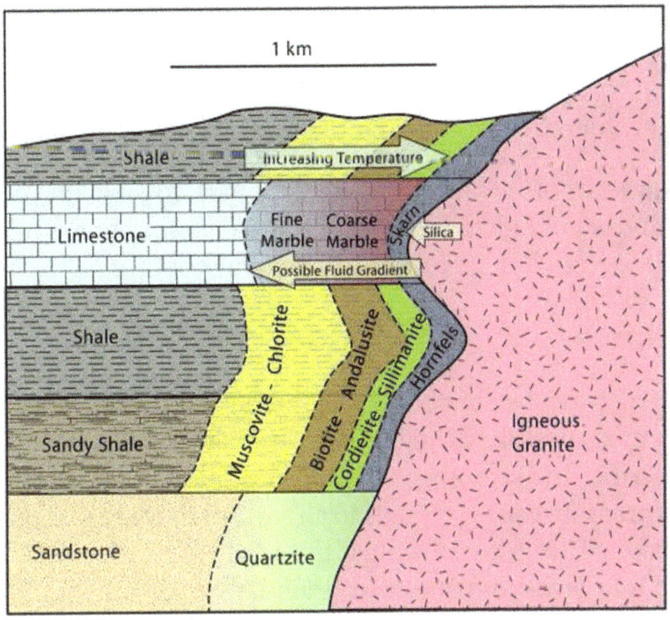

Contact metamorphism and associated minerals for different types of country rock. Aluminium-rich shale will form indicator minerals such as cordierite and sillimanite. Impure limestone may host indicator minerals such as olivine. Where there are significant volumes of fluid present, skarns may form.

Metamorphic Minerals

All rocks are composed of minerals, and each mineral is only stable over a particular range of temperatures and pressures. Outwith that range it will tend to break down or combine with a neighbouring mineral to form a new mineral.

The process of metamorphism is associated with the formation of particular metamorphic minerals. Some silicate minerals are so diagnostic of metamorphism that their presence indicates that the rock is metamorphic. These mineral are known as index minerals and include andalusite, kyanite, staurolite and sillimanite. Some other minerals, such as garnet, muscovite, feldspar, quartz and micas, are also present in metamorphic rocks, but these may also be found in igneous and sedimentary rocks.

Metamorphic Zones

The concept of metamorphic zones can be illustrated by considering what happens to mudrock as it becomes subjected to increasing levels of metamorphism. Mudrock is a fine-grained sedimentary rock which often contains aluminium-rich minerals, and it produces the following minerals as it is progressively metamorphosed, from a low-grade to a high-grade rock:

- Chlorite zone: Quartz, chlorite, muscovite, albite.
- Biotite zone: Quartz, muscovite, biotite, chlorite, albite.
- Garnet zone: Quartz, muscovite, biotite, garnet, Na plagioclase.
- Staurolite zone: Quartz, muscovite, biotite, garnet, staurolite, plagioclase.
- Kyanite zone: Quartz, muscovite, biotite, garnet, kyanite, plagioclase, +/- staurolite.
- Sillimanite zone: Quartz, muscovite, biotite, garnet, sillimanite, plagioclase.

In extreme conditions, metamorphic rocks can become partially melted. When the partially melted rock cools, it is composed of light bands consisting of igneous components and dark bands of unmelted metamorphic material. Such rocks are called migmatites.

Metamorphic Textures

Just as a metamorphic rock can be defined by its mineral assemblage, it can also be defined by its texture. This may be a more useful method of categorizing metamorphic rocks when in the field. The term "texture" describes the size, shape and arrangement of grains within a rock. During regional metamorphism, pressure causes crystals to align until they are at right angles to the direction of maximum compression. This occurs through the gradual rotation of existing crystals and the growth of new crystals in a pressure-controlled orientation. Alignment of minerals into planes is called foliation. As a general rule, the coarser the grain size of the rock, the higher its metamorphic grade. For example, it may be impossible to identify individual minerals in a slate with a hand lens; in schist it may be possible to differentiate minerals even with the naked eye, and certainly with a hand lens.

Foliated Textures

- Slaty cleavage: Closely spaced planar surfaces along which rocks are easily split; seen in slate, which is a very fine-grained foliated rock derived from an original shale protolith.
- Schistosity: Medium sized platy minerals in a preferred orientation, often interlayered with quartz and feldspar; characteristic of the rock type schist.
- Gneissic texture: Seen in high-grade metamorphic rocks and characterised by medium- to coarse- grained crystals arranged in alternating bands of dark and light minerals (gneissic banding).

Non-foliated Textures

- Granoblastic: A mosaic of equidimensional grains; hornfels, marble and quartzite have a granoblastic texture.
- Porphyroblastic: Characterised by porphyroblasts, which are large crystals surrounded by a finer-grained groundmass in a metamorphic rock. This texture is often seen in garnet-mica schists.

Slate (left) & phyllite. (Phyllite is intermediate in metamorphic grade between slate and schist).

Garnet-mica schist.

Garnet-biotite gneiss.

Migmatite.

Magma

A magma consists mostly of liquid rock matter, but may contain crystals of various minerals, and may contain a gas phase that may be dissolved in the liquid or may be present as a separate gas phase.

Magma can cool to form an igneous rock either on the surface of the Earth - in which case it produces a volcanic or extrusive igneous rock, or beneath the surface of the Earth, - in which case it produces a plutonic or intrusive igneous rock.

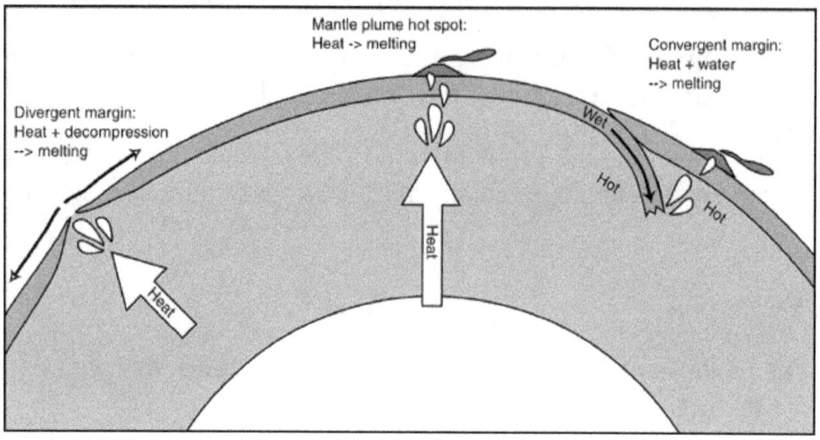

Sources of Magma.

Partial Melting

Melting of solid rocks to form magma is controlled by three physical parameters: its temperature, pressure, and composition. Mechanisms are discussed in the entry for igneous rock.

When rocks melt they do so incrementally and gradually; most rocks are made of several minerals, all of which have different melting points, and the physical/chemical relationships controlling melting are complex. As a rock melts, its volume changes. When enough rock is melted, the small globules of melt (generally occurring in between mineral grains) link up and soften the rock. Under pressure within the earth, as little as a fraction of a percent partial melting may be sufficient to cause melt to be squeezed from its source. Melts can stay in place long enough to melt to 20% or even 35%, but rocks are rarely melted in excess of 50%, because eventually the melted rock mass becomes a crystal and melt mush that can then ascend en masse as a diapir, which may then cause further decompression melting.

Geochemical Implications of Partial Melting

The degree of partial melting is critical for determining what type of magma is produced. The degree of partial melting required to form a melt can be estimated by considering the relative enrichment of incompatible elements versus compatible elements. Incompatible elements commonly include potassium, barium, caesium, rubidium.

Rock types produced by small degrees of partial melting in the Earth's mantle are typically alkaline (Ca, Na), potassic (K) and peralkaline (high aluminium to silica ratio). Typically, primitive melts of this composition form lamprophyre, lamproite, kimberlite and sometimes nepheline-bearing mafic rocks such as alkali basalts and essexite gabbros or even carbonatite.

Pegmatite may be produced by low degrees of partial melting of the crust. Some granite-composition magmas are eutectic (or cotectic) melts, and they may be produced by low to high degrees of partial melting of the crust, as well as by fractional crystallization. At high degrees of partial melting of the crust, granitoids such as tonalite, granodiorite and monzonite can be produced, but other mechanisms are typically important in producing them.

Magma usage for Energy Production

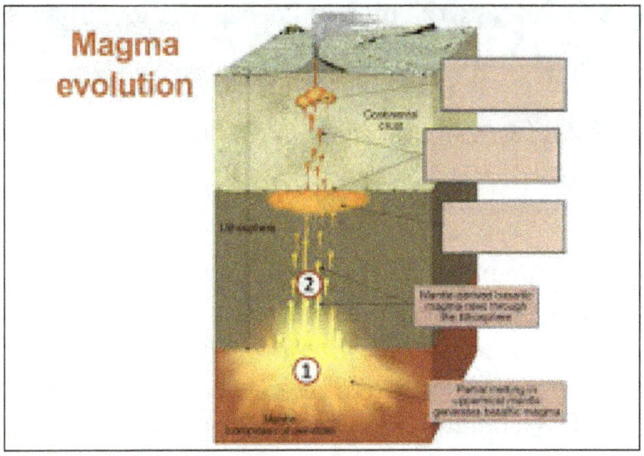

Evolution of Magmas.

The Iceland Deep Drilling Project, while drilling several 5000m holes in an attempt to harness the heat in the volcanic bedrock below the surface of Iceland, struck a pocket of magma at 2,100m. Being only the third time in recorded history that magma had been reached, IDDP decided to invest in the hole, naming it IDDP-1.

A cemented steelcase was constructed in the hole with a perforation at the bottom close to the magma. The high temperatures and pressure of the magma steam were used to generate 36MW of power, making IDDP-1 the world's first magma-enhanced geothermal system.

Primary Melts

When a rock melts, the liquid is a primary melt. Primary melts have not undergone any differentiation and represent the starting composition of a magma. In nature it is rare to find primary melts. The leucosomes of migmatites are examples of primary melts. Primary melts derived from the mantle are especially important, and are known as primitive melts or primitive magmas. By finding the primitive magma composition of a magma series it is possible to model the composition of the mantle from which a melt was formed, which is important in understanding evolution of the mantle.

Parental Melts

Where it is impossible to find the primitive or primary magma composition, it is often useful to attempt to identify a parental melt. A parental melt is a magma composition from which the observed range of magma chemistries has been derived by the processes of igneous differentiation. It need not be a primitive melt.

For instance, a series of basalt flows are assumed to be related to one another. A composition from which they could reasonably be produced by fractional crystallization is termed a parental melt. Fractional crystallization models would be produced to test the hypothesis that they share a common parental melt.

Characteristics of Magma

Types of magma are determined by chemical composition of the magma. Three general types are recognized:

- Basaltic magma -- SiO_2 45-55 wt%, high in Fe, Mg, Ca, low in K, Na.
- Andesitic magma -- SiO_2 55-65 wt%, intermediate. in Fe, Mg, Ca, Na, K.
- Rhyolitic magma -- SiO_2 65-75%, low in Fe, Mg, Ca, high in K, Na.

Gases in Magmas

At depth in the Earth nearly all magmas contain gas dissolved in the liquid, but the gas forms a separate vapor phase when pressure is decreased as magma rises toward the surface. This is similar to carbonated beverages which are bottled at high pressure. The high pressure keeps the gas in solution in the liquid, but when pressure is decreased, like when you open the can or bottle, the gas comes out of solution and forms a separate gas phase that you see as bubbles. Gas gives magmas their explosive character, because volume of gas expands as pressure is reduced. The composition of the gases in magma are:

- Mostly H_2O (water vapor) with some CO_2 (carbon dioxide).
- Minor amounts of Sulfur, Chlorine, and Fluorine gases.

The amount of gas in a magma is also related to the chemical composition of the magma. Rhyolitic magmas usually have higher dissolved gas contents than basaltic magmas.

Temperature of Magmas

Temperature of magmas is difficult to measure (due to the danger involved), but laboratory measurement and limited field observation indicate that the eruption temperature of various magmas is as follows:

- Basaltic magma - 1000 to 1200 °C.
- Andesitic magma - 800 to 1000 °C.
- Rhyolitic magma - 650 to 800 °C.

Viscosity of Magmas

Viscosity is the resistance to flow (opposite of fluidity). Viscosity depends on primarily on the composition of the magma, and temperature.

- Higher SiO_2 (silica) content magmas have higher viscosity than lower SiO_2 content magmas (viscosity increases with increasing SiO_2 concentration in the magma).
- Lower temperature magmas have higher viscosity than higher temperature magmas (viscosity decreases with increasing temperature of the magma).

Thus, basaltic magmas tend to be fairly fluid (low viscosity), but their viscosity is still 10,000 to

100,0000 times more viscous than water. Rhyolitic magmas tend to have even higher viscosity, ranging between 1 million and 100 million times more viscous than water. Viscosity is an important property in determining the eruptive behavior of magmas.

Table					
Magma Type	Solidified Rock	Chemical Composition	Temperature	Viscosity	Gas Content
Basaltic	Basalt	45-55 SiO_2 %, high in Fe, Mg, Ca, low in K, Na	1000 - 1200 °C	$10 - 10^3$ PaS	Low
Andesitic	Andesite	55-65 SiO_2 %, intermediate in Fe, Mg, Ca, Na, K	800 - 1000 °C	$10^3 - 10^5$ PaS	Intermediate
Rhyolitic	Rhyolite	65-75 SiO_2 %, low in Fe, Mg, Ca, high in K, Na.	650 - 800 °C	$10^5 - 10^9$ PaS	High

Lava

10-metre-high (33 ft) fountain of pāhoehoe lava, Hawaii, United States.

Lava is molten rock generated by geothermal energy and expelled through fractures in planetary crust or in an eruption, usually at temperatures from 700 to 1,200 °C (1,292 to 2,192 °F). The structures resulting from subsequent solidification and cooling are also sometimes described as *lava*. The molten rock is formed in the interior of some planets, including Earth, and some of their satellites, though such material located below the crust is referred to by other terms.

Lava flow during a rift eruption at Krafla, Iceland.

A lava flow is a moving outpouring of lava created during a non-explosive effusive eruption. When it has stopped moving, lava solidifies to form igneous rock. The term *lava flow* is commonly shortened to *lava*. Although lava can be up to 100,000 times more viscous than water, lava can flow great distances before cooling and solidifying because of its thixotropic and shear thinning properties.

Explosive eruptions produce a mixture of volcanic ash and other fragments called tephra, rather than lava flows. The first use in connection with extruded magma (molten rock below the Earth's surface) was apparently in a short account written by Francesco Serao on the eruption of Vesuvi-usin 1737. Serao described "a flow of fiery lava" as an analogy to the flow of water and mud down the flanks of the volcano following heavy rain.

Lava Composition

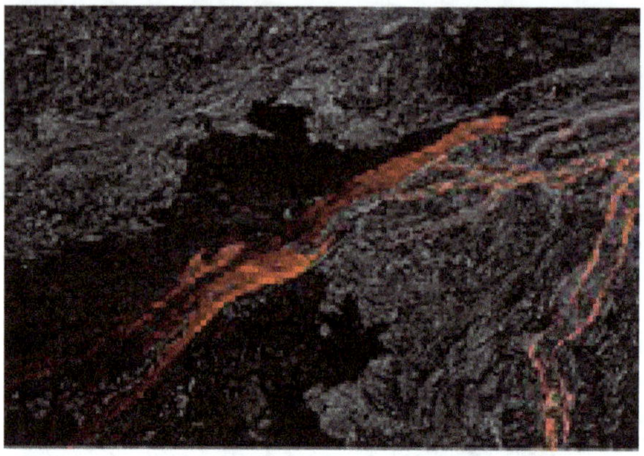

Pāhoehoe and ʻaʻā lava flows side by side in Hawaii.

The composition of almost all lava of the Earth's crust is dominated by silicate minerals, mostly feldspars, olivine, pyroxenes, amphiboles, micas and quartz.

Silicate Lavas

Igneous rocks, which form lava flows when erupted, can be classified into three chemical types: *felsic*, *intermediate*, and *mafic* (four if one includes the super-heated*ultramafic*). These classes are primarily chemical, however, the chemistry of lava also tends to correlate with the magma temperature, its viscosity and its mode of eruption.

Felsic Lava

Felsic or silicic lavas such as rhyolite and dacite typically form lava spines, lava domes or "coulees" (which are thick, short lava flows) and are associated with pyroclastic (fragmental) deposits. Most silicic lava flows are extremely viscous, and typically fragment as they extrude, producing blocky autobreccias. The high viscosity and strength are the result of their chemistry, which is high in silica, aluminium, potassium, sodium, and calcium, forming a polymerized liquid rich in feldspar and quartz, and thus has a higher viscosity than other magma types. Felsic magmas can erupt at temperatures as low as 650 to 750 °C (1,202 to 1,382 °F). Unusually hot (>950 °C; >1,740 °F) rhyolite lavas, however, may flow for distances of many tens of kilometres, such as in the Snake River Plain of the northwestern United States.

Intermediate Lava

Intermediate or andesitic lavas are lower in aluminium and silica, and usually somewhat richer in magnesium and iron. Intermediate lavas form andesite domes and block lavas, and may occur on steep composite volcanoes, such as in the Andes. Poorer in aluminium and silica than felsic lavas, and also commonly hotter (in the range of 750 to 950 °C (1,380 to 1,740 °F)), they tend to be less viscous. Greater temperatures tend to destroy polymerized bonds within the magma, promoting more fluid behaviour and also a greater tendency to form phenocrysts. Higher iron and magnesium tends to manifest as a darker groundmass, and also occasionally amphibole or pyroxene phenocrysts.

Mafic Lava

Mafic or basaltic lavas are typified by their high ferromagnesian content, and generally erupt at temperatures in excess of 950 °C (1,740 °F). Basaltic magma is high in iron and magnesium, and has relatively lower aluminium and silica, which taken together reduces the degree of polymerization within the melt. Owing to the higher temperatures, viscosities can be relatively low, although still thousands of times higher than water. The low degree of polymerization and high temperature favors chemical diffusion, so it is common to see large, well-formed phenocrysts within mafic lavas. Basalt lavas tend to produce low-profile shield volcanoes or "flood basalt fields", because the fluidal lava flows for long distances from the vent. The thickness of a basalt lava, particularly on a low slope, may be much greater than the thickness of the moving lava flow at any one time, because basalt lavas may "inflate" by supply of lava beneath a solidified crust. Most basalt lavas are of *'A'ā* or *pāhoehoe* types, rather than block lavas. Underwater, they can form pillow lavas, which are rather similar to entrail-type pahoehoe lavas on land.

Ultramafic Lava

Ultramafic lavas such as komatiite and highly magnesian magmas that form boninite take the composition and temperatures of eruptions to the extreme. Komatiites contain over 18% magnesium oxide, and are thought to have erupted at temperatures of 1,600 °C (2,910 °F). At this temperature there is no polymerization of the mineral compounds, creating a highly mobile liquid. Most if not all ultramafic lavas are no younger than the Proterozoic, with a few ultramafic magmas known from the Phanerozoic. No modern komatiite lavas are known, as the Earth's mantle has cooled too much to produce highly magnesian magmas.

Unusual Lavas

Some lavas of unusual composition have erupted onto the surface of the Earth. These include:

- Carbonatite and natrocarbonatite lavas are known from Ol Doinyo Lengai volcano in Tanzania, which is the sole example of an active carbonatite volcano.
- Iron oxide lavas are thought to be the source of the iron ore at Kiruna, Sweden which formed during the Proterozoic.Iron oxide lavas of Pliocene age occur at the El Laco volcanic complex on the Chile-Argentina border. Iron oxide lavas are thought to be the result of immiscible separation of iron oxide magma from a parental magma of calc-alkaline or alkaline composition.

- Sulfur lava flows up to 250 metres (820 feet) long and 10 metres (33 feet) wide occur at Lastarria volcano, Chile. They were formed by the melting of sulfur deposits at temperatures as low as 113 °C (235 °F).
- Olivine nephelinite lavas are thought to have come from much deeper in the mantle of the Earth than other lavas.

The term "lava" can also be used to refer to molten "ice mixtures" in eruptions on the icy satellites of the Solar System's gas giants.

Lava Behavior

Toes of a pāhoehoe advance across a road in Kalapana on the east rift zone of Kīlauea Volcano in Hawaii, United States.

In general, the composition of a lava determines its behavior more than the temperature of its eruption. The viscosity of lava is important because it determines how the lava will behave. Lavas with high viscosity are rhyolite, dacite, andesite and trachyte, with cooled basaltic lava also quite viscous; those with low viscosities are freshly erupted basalt, carbonatite and occasionally andesite.

Highly viscous lava shows the following behaviors:

- Tends to flow slowly, clog, and form semi-solid blocks which resist flow.
- Tends to entrap gas, which form vesicles (bubbles) within the rock as they rise to the surface.
- Correlates with explosive or phreatic eruptions and is associated with tuff and pyroclastic-flows.

Highly viscous lavas do not usually flow as liquid, and usually form explosive fragmental ash or tephra deposits. However, a degassed viscous lava or one which erupts somewhat hotter than usual may form a lava flow.

Lava with low viscosity shows the following behaviors:

- Tends to flow easily, forming puddles, channels, and rivers of molten rock.
- Tends to easily release bubbling gases as they are formed.

- Eruptions are rarely pyroclastic and are usually quiescent.
- Volcanoes tend to form broad shields rather than steep cones.

Lavas also may contain many other components, sometimes including solid crystals of various minerals, fragments of exotic rocks known as xenoliths and fragments of previously solidified lava.

Lava flow speeds vary based primarily on viscosity and slope. In general, lava flows slowly (0.25 mph), with maximum speeds between 6–30 mph on steep slopes. An exceptional speed of 20–60 mph was recorded following the collapse of a lava lake at Mount Nyiragongo.

Lava Morphology

Lava entering the sea to expand the big island of Hawaii, Hawaii Volcanoes National Park.

The physical behavior of lava creates the physical forms of a lava flow or volcano. More fluid basaltic lava flows tend to form flat sheet-like bodies, whereas viscous rhyolite lava flows forms knobbly, blocky masses of rock.

General features of volcanology can be used to classify volcanic edifices and provide information on the eruptions which formed the lava flow, even if the sequence of lavas have been buried or metamorphosed.

The ideal lava flow will have a brecciatedtop, either as pillow lava development, autobreccia and rubble typical of *ʻaʻā* and viscous flows, or a vesicular or frothy carapace such as scoria or pumice. The top of the lava will tend to be glassy, having been flash frozen in contact with the air or water.

The centre of a lava flow is commonly massive and crystalline, flow banded or layered, with microscopic groundmass crystals. The more viscous lava forms tend to show sheeted flow features, and blocks or breccia entrained within the sticky lava. The crystal size at the centre of a lava will in general be greater than at the margins, as the crystals have more time to grow.

The base of a lava flow may show evidence of hydrothermal activity if the lava flowed across moist or wet substrates. The lower part of the lava may have vesicles, perhaps filled with minerals (amygdules). The substrate upon which the lava has flowed may show signs of scouring, it may be broken or disturbed by the boiling of trapped water, and in the case of soil profiles, may be baked into a brick-red terracotta.

Lava enters the Pacific at the Big Island of Hawaii.

Discriminating between an intrusive sill and a lava flow in ancient rock sequences can be difficult. However, some sills do not usually have brecciated margins, and may show a weak metamorphic aureole on both the upper and lower surface, whereas a lava will only bake the substrate beneath it. However, it is often difficult in practice to identify these metamorphic phenomena because they are usually weak and restricted in size. Peperitic sills, intruded into wet sedimentary rocks, commonly do not bake upper margins and have upper and lower autobreccias, closely similar to lavas.

ʻAʻā

Glowing ʻaʻā flow front advancing over pāhoehoe on the coastal plain of Kilauea in Hawaii, United States.

ʻAʻā is one of three basic types of flow lava. ʻAʻā is basaltic lava characterized by a rough or rubbly surface composed of broken lava blocks called clinker. The Hawaiian word was introduced as a technical term in geology by Clarence Dutton.

The loose, broken, and sharp, spiny surface of an ʻaʻā flow makes hiking difficult and slow. The clinkery surface actually covers a massive dense core, which is the most active part of the flow. As pasty lava in the core travels downslope, the clinkers are carried along at the surface. At the leading edge of an ʻaʻā flow, however, these cooled fragments tumble down the steep front and are buried by the advancing flow. This produces a layer of lava fragments both at the bottom and top of an ʻaʻā flow.

Accretionary lava balls as large as 3 metres (10 feet) are common on ʻaʻā flows. ʻAʻā is usually of higher viscosity than pāhoehoe. Pāhoehoe can turn into ʻaʻā if it becomes turbulent from meeting impediments or steep slopes.

The sharp, angled texture makes ‘a‘ā a strong radar reflector, and can easily be seen from an orbiting satellite (bright on Magellan pictures).

‘A‘ā lavas typically erupt at temperatures of 1,000 to 1,100 °C (1,830 to 2,010 °F).

Pāhoehoe

Pāhoehoe ,also spelled *pahoehoe,* is basaltic lava that has a smooth, billowy, undulating, or ropy surface. These surface features are due to the movement of very fluid lava under a congealing surface crust. The Hawaiian word was introduced as a technical term in geology by Clarence Dutton.

A pāhoehoe flow typically advances as a series of small lobes and toes that continually break out from a cooled crust. It also forms lava tubes where the minimal heat loss maintains low viscosity. The surface texture of pāhoehoe flows varies widely, displaying all kinds of bizarre shapes often referred to as lava sculpture. With increasing distance from the source, pāhoehoe flows may change into ‘a‘ā flows in response to heat loss and consequent increase in viscosity. Pahoehoe lavas typically have a temperature of 1,100 to 1,200 °C (2,010 to 2,190 °F).

Pāhoehoe lava from Kīlauea volcano,
Hawaii, United States.

On the Earth, most lava flows are less than 10 km (6.2 mi) long, but some pāhoehoe flows are more than 50 km (31 mi) long.

The rounded texture makes pāhoehoe a poor radar reflector, and is difficult to see from an orbiting satellite (dark on Magellan picture).

Block Lava Flows

Block lava flows are typical of andesitic lavas from stratovolcanoes. They behave in a similar manner to ‘a‘ā flows but their more viscous nature causes the surface to be covered in smooth-sided angular fragments (blocks) of solidified lava instead of clinkers. Like in ‘a‘ā flows, the molten interior of the flow, which is kept insulated by the solidified blocky surface, overrides the rubble that falls off the flow front. They also move much more slowly downhill and are thicker in depth than ‘a‘ā flows.

Domes and Coulées

Lava domes and coulées are associated with felsic lava flows ranging from dacite to rhyolite. The very viscous nature of these lava cause them to not flow far from the vent, causing the lava to form a lava dome at the vent. When a dome forms on an inclined surface it can flow in short thick flows called coulées (dome flows). These flows often travel only a few kilometers from the vent.

Pillow Lava

Pillow lava is the lava structure typically formed when lava emerges from an underwater volcanic vent or subglacial volcano or a lava flow enters the ocean. However, pillow lava can also form when lava is erupted beneath thick glacial ice. The viscous lava gains a solid crust on contact with the water, and this crust cracks and oozes additional large blobs or "pillows" as more lava emerges from the advancing flow. Since water covers the majority of Earth's surface and most volcanoes are situated near or under bodies of water, pillow lava is very common.

Pillow lava.

Lava Landforms

Because it is formed from viscous molten rock, lava flows and eruptions create distinctive formations, landforms and topographical features from the macroscopic to the microscopic.

Volcanoes

Arenal Volcano, Costa Rica, is a stratovolcano.

Volcanoes are the primary landforms built by repeated eruptions of lava and ash over time. They range in shape from shield volcanoes with broad, shallow slopes formed from predominantly effusive eruptions of relatively fluid basaltic lava flows, to steeply-sided stratovolcanoes (also known as composite volcanoes) made of alternating layers of ash and more viscous lava flows typical of intermediate and felsic lavas.

A caldera, which is a large subsidence crater, can form in a stratovolcano, if the magma chamber is partially or wholly emptied by large explosive eruptions; the summit cone no longer supports itself and thus collapses in on itself afterwards. Such features may include volcanic crater lakes and lava domes after the event. However, calderas can also form by non-explosive means such as gradual magma subsidence. This is typical of many shield volcanoes.

Cinder and Spatter Cones

Cinder cones and spatter cones are small-scale features formed by lava accumulation around a small vent on a volcanic edifice. Cinder cones are formed from tephra or ash and tuff which is thrown from an explosive vent. Spatter cones are formed by accumulation of molten volcanic slag and cinders ejected in a more liquid form.

Kīpukas

Another Hawaiian English term derived from the Hawaiian language, a kīpuka denotes an elevated area such as a hill, ridge or old lava dome inside or downslope from an area of active volcanism. New lava flows will cover the surrounding land, isolating the kīpuka so that it appears as a (usually) forested island in a barren lava flow.

Lava Domes

A forested lava dome in the midst of the Valle Grande, the largest meadow in the Valles Caldera National Preserve, New Mexico, United States.

Lava domes are formed by the extrusion of viscous felsic magma. They can form prominent rounded protuberances, such as at Valles Caldera. As a volcano extrudes silicic lava, it can form an *inflation dome*, gradually building up a large, pillow-like structure which cracks, fissures, and may release cooled chunks of rock and rubble. The top and side margins of an inflating lava dome tend to be covered in fragments of rock, breccia and ash.

Examples of lava dome eruptions include the Novarupta dome, and successive lava domes of Mount St Helens.

Lava Tubes

Lava tubes are formed when a flow of relatively fluid lava cools on the upper surface sufficiently to form a crust. Beneath this crust, which being made of rock is an excellent insulator, the lava can continue to flow as a liquid. When this flow occurs over a prolonged period of time the lava conduit can form a tunnel-like aperture or *lava tube*, which can conduct molten rock many kilometres from the vent without cooling appreciably. Often these lava tubes drain out once the supply of fresh lava has stopped, leaving a considerable length of open tunnel within the lava flow.

Lava tubes are known from the modern day eruptions of Kīlauea, and significant, extensive and open lava tubes of Tertiary age are known from North Queensland, Australia, some extending for 15 kilometres (9 miles).

Lava Fountains

A lava fountain is a volcanic phenomenon in which lava is forcefully but non-explosively ejected from a crater, vent, or fissure. The highest lava fountains recorded were during the 1999 eruption of Mount Etna in Italy, which reached heights of 2,000 m (6,562 ft).However, lava fountains observed during Mount Vesuvius' 1779 eruption are believed to have reached at least 3,000 m (9,843 ft). Lava fountains may occur as a series of short pulses, or a continuous jet of lava. They are commonly associated with Hawaiian eruptions.

450m-high lava fountain at Kilauea.

Lava Lakes

Rarely, a volcanic cone may fill with lava but not erupt. Lava which pools within the caldera is known as a lava lake. Lava lakes do not usually persist for long, either draining back into the magma chamber once pressure is relieved (usually by venting of gases through the caldera), or by draining via eruption of lava flows or pyroclastic explosion.

There are only a few sites in the world where permanent lakes of lava exist. These include:

- Mount Erebus, Antarctica.
- Pu'u 'Ō'ō and Halemaumau Crater on Kilauea volcano, Hawaii.

- Erta Ale, Ethiopia.
- Nyiragongo, Democratic Republic of Congo.
- Ambrym, Vanuatu.

Shiprock, New Mexico, United States: a volcanic neck in the distance, with a radiating dike on its south side.

Lava Delta

Lava deltas form wherever sub-aerial flows of lava enter standing bodies of water. The lava cools and breaks up as it encounters the water, with the resulting fragments filling in the seabed topography such that the sub-aerial flow can move further offshore. Lava deltas are generally associated with large-scale, effusive type basaltic volcanism.

Hazards

Lava flows are enormously destructive to property in their path. However, casualties are rare since flows are usually slow enough for people and animals to escape, though this is dependent on the viscosity of the lava. Nevertheless, injuries and deaths have occurred, either because they had their escape route cut off, because they got too close to the flow or, more rarely, if the lava flow front travels too quickly. This notably happened during the eruption of Nyiragongo in Zaire (now Democratic Republic of the Congo). On the night of 10 January 1977 a crater wall was breached and a fluid lava lake drained out in under an hour. The resulting flow sped down the steep slopes at up to 100 km/h (62 mph), and overwhelmed several villages while residents were asleep. As a result of this disaster, the mountain was designated a Decade Volcano in 1991.

Deaths attributed to volcanoes frequently have a different cause, for example volcanic ejecta, pyroclastic flow from a collapsing lava dome, lahars, poisonous gases that travel ahead of lava, or explosions caused when the flow comes into contact with water. A particularly dangerous area is called a lava bench. This very young ground will typically break-off and fall into the sea.

Areas of recent lava flows continue to represent a hazard long after the lava has cooled. Where young flows have created new lands, land is more unstable and can break-off into the sea. Flows often crack deeply, forming dangerous chasms, and a fall against 'a'a lava is similar to falling against broken glass. Rugged hiking boots, long pants, and gloves are recommended when crossing lava flows.

Diverting a lava flow is extremely difficult, but it can be accomplished in some circumstances, as was once partially achieved in Vestmannaeyjar, Iceland.

References

- Rock-geology: britannica.com, Retrieved 23 August, 2019
- Formation-types-and-examples-of-igneous-rocks: eartheclipse.com, Retrieved 02 June, 2019
- About-sedimentary-rocks-1438951: thoughtco.com, Retrieved 25 April, 2019
- Metamorphic-rocks: geologyglasgow.org.uk, Retrieved 03 March, 2019
- Magma-characteristics-types-sources-and: geologyin.com, Retrieved 15 January, 2019

Principles and Laws of Geology

3

- Stratigraphic Laws
- Principle of Original Horizontality
- Cross-cutting Relationships
- Uniformitarianism
- Law of Superposition
- Principle of Lateral Continuity
- Law of Included Fragments
- Principle of Faunal Succession

There are various laws and principles which are studied within the discipline of geology. A few of them are the principle of original horizontality, principle of lateral continuity, principle of faunal succession, law of included fragments and law of superposition. This chapter discusses in detail these principles and laws of geology.

Stratigraphic Laws

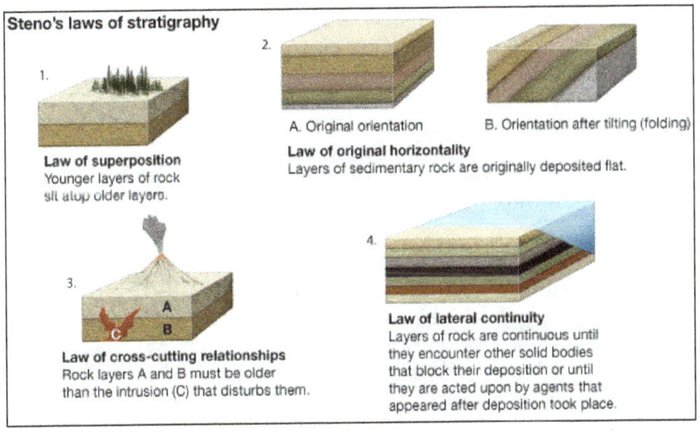

Stratigraphic Laws are basic principles that all geologists use in deciphering the spatial and temporal relationships of rock layers. These laws were developed in the 17th to 19th centuries based upon the work of Niels Steno, James Hutton and William Smith, among others.

Principle of Original Horizontality

The Principle of Original Horizontality was proposed by the Danish geological pioneer Nicholas Steno. This principle states that layers of sediment are originally deposited horizontally under the action of gravity. The principle is important to the analysis of folded and tilted strata.

From these observations is derived the conclusion that the Earth has not been static and that great forces have been at work over long periods of time, further leading to the conclusions of the science of plate tectonics; that movement and collisions of large plates of the Earth's crust is the cause of folded strata.

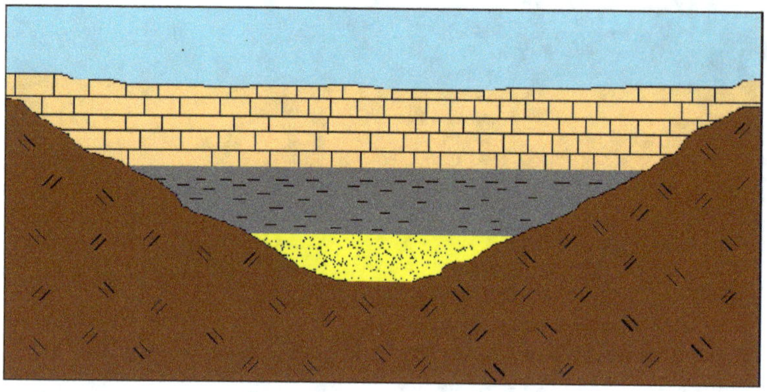

Original horizontal strata.

As one of Steno's Laws, the Principle of Original Horizontality served well in the nascent days of geological science. However, it is now known that not all sedimentary layers are deposited purely horizontally.

For instance, coarser grained sediments such as sand may be deposited at angles of up to 15 degrees, held up by the internal friction between grains which prevents them slumping to a lower angle without additional reworking or effort. This is known as the angle of repose, and a prime example is the surface of sand dunes.

Cross-cutting Relationships

Cross-cutting relationships is a principle of geology that states that the geologic feature which cuts another is the younger of the two features. It is a relative dating technique in geology. It was first developed by Danish geological pioneer Nicholas Steno in *Dissertationis prodromus* and later formulated by James Hutton in *Theory of the Earth* and embellished upon by Charles Lyell in *Principles of Geology*.

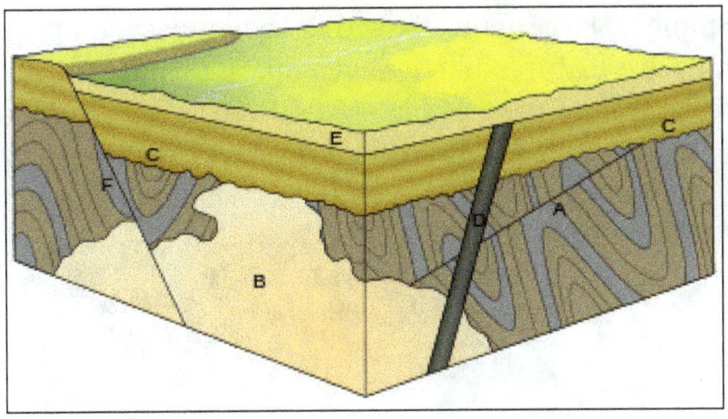

Cross-cutting relations can be used to determine the relative ages of rock strata and other geological structures. Explanations: A – folded rock strata cut by a thrust fault; B – large intrusion (cutting through A); C – erosional angular unconformity (cutting off A & B) on which rock strata were deposited; D – volcanic dike (cutting through A, B & C); E – even younger rock strata (overlying C & D); F – normal fault(cutting through A, B, C & possibly E).

Types

There are several basic types of cross cutting relationships:

- Structural relationships may be faults or fractures cutting through an older rock.
- Intrusional relationships occur when an igneous pluton or dike is intruded into pre-existing rocks.
- Stratigraphic relationships may be an erosional surface (or unconformity) cuts across older rock layers, geological structures, or other geological features.
- Sedimentological relationships occur where currents have eroded or scoured older sediment in a local area to produce, for example, a channel filled with sand.
- Paleontological relationships occur where animal activity or plant growth produces truncation. This happens, for example, where animal burrows penetrate into pre-existing sedimentary deposits.
- Geomorphological relationships may occur where a surficial feature, such as a river, flows through a gap in a ridge of rock. In a similar example, an impact crater excavates into a subsurface layer of rock.

Cross-cutting relationships may be compound in nature. For example, if a fault were truncated by an unconformity, and that unconformity cut by a dike. Based upon such compound cross-cutting relationships it can be seen that the fault is older than the unconformity which in turn is older than the dike. Using such rationale, the sequence of geological events can be better understood.

Scale

Cross-cutting relationships may be seen cartographically, megascopically, and microscopically. In other words, these relationships have various scales. A cartographic crosscutting relationship might look like, for example, a large fault dissecting the landscape on a large map. Megascopic cross-cutting relationships are features like igneous dikes, as mentioned above, which would be seen on an outcrop or in a limited geographic area. Microscopic cross-cutting relationships are those that require study by magnification or other close scrutiny. For example, penetration of a fossil shell by the drilling action of a boring organism is an example of such a relationship.

Cross-cutting relationships involving an andesitic dike in Peru that cuts across the lower sedimentary strata. Both the dike and the lower strata are cut by an unconformity.

A light-gray igneous intrusion in Sweden cut by a younger white pegmatite dike, which in turn is cut by an even younger black diabase dike.

Other Use

Cross-cutting relationships can also be used in conjunction with radiometric age dating to effect an age bracket for geological materials that cannot be directly dated by radiometric techniques. For example, if a layer of sediment containing a fossil of interest is bounded on the top and bottom by unconformities, where the lower unconformity truncates dike A and the upper unconformity truncates dike B (which penetrates the layer in question), this method can be used. A radiometric age date from crystals in dike A will give the maximum age date for the layer in question and likewise, crystals from dike B will give us the minimum age date.

Uniformitarianism

Uniformitarianism, in geology, the doctrine suggesting that Earth's geologic processes acted in the same manner and with essentially the same intensity in the past as they do in the present and that such uniformity is sufficient to account for all geologic change. This principle is fundamental to geologic thinking and underlies the whole development of the science of geology.

When William Whewell, a University of Cambridge scholar, introduced the term in 1832, the prevailing view (called catastrophism) was that Earth had originated through supernatural means and had been affected by a series of catastrophic events such as the biblical Flood. In contrast to catastrophism, uniformitarianism postulates that phenomena displayed in rocks may be entirely accounted for by geologic processes that continue to operate—in other words, the present is the key to the past.

The expression uniformitarianism, however, has passed into history, because the argument between catastrophists and uniformitarians has largely died. Geology as an applied science draws on the other sciences, but in the early 19th century, geologic discovery had outrun the physics and chemistry of the day. As geologic phenomena became understandable in terms of advancing physics, chemistry, and biology, the reality of the principle of uniformity as a major philosophical tenet of geology became established, and the controversy between catastrophists and uniformitarians largely ended.

Hutton's Contributions

The idea that the laws that govern geologic processes have not changed during Earth's history was first expressed by Scottish geologist James Hutton, who in 1785 presented his ideas—later published in two volumes as Theory of the Earth—at meetings of the Royal Society of Edinburgh. Hutton showed that Earth had a long history that could be interpreted in terms of processes observed in the present. He showed, for instance, how soils were formed by the weathering of rocks and how layers of sediment accumulated on Earth's surface.

He also stated that there was no need of any preternatural cause to explain the geologic record. Hutton's proposal challenged the concept of a biblical Earth (with a history of some 6,000 years) that was created especially to be a home for human beings; the effect of his ideas on the learned world can be compared only the earlier revolution in thought brought about by Polish astronomer Nicolaus Copernicus, German astronomer Johannes Kepler, and Italian astronomer Galileo when they displaced the concept of a universe centred on Earth with the concept of a solar system centred on the Sun. Both advances challenged existing thought and were fiercely resisted for many years.

Lord Kelvin's Contributions

The publication in 1859 of the conclusions of Darwin and British naturalist Alfred Russel Wallace on the origin of species extended the principle of uniformity to the plant and animal kingdoms. Although catastrophists continued to fight a rearguard action against the Huttonian-Lyellian-Darwinian view until the end of the 19th century, a new criticism was raised by William Thomson (later Lord Kelvin), one of the leading researchers on thermodynamics. Thomson pointed out that Earth loses heat by thermal conduction and that geologic processes may have changed as a consequence; he also concluded that this cooling placed an upper limit on Earth's age. With the discovery of radioactivity and the recognition that radioactive isotopes within the planet provide a continuing internal source of heat, it became clear that Thomson's conclusion that Earth was less than 100 million years old was incorrect, but his argument that Earth suffers an irreversible loss of energy remains valid.

The heat loss, owing in part to the decay of the heat-producing radioactive isotopes (atoms such as uranium-235, uranium-238, and thorium-232), has an important consequence. Although the principle of uniformity is correct in that physical laws have not changed over geologic time, Earth's behaviour has changed as temperatures have fallen, with the consequence that the extent of igneous activity and movement of Earth's crust has changed during geologic time. Thus, it is possible that the plate tectonics that operates today and that has operated in past geologic intervals of time was preceded by somewhat different processes of deformation during Precambrian time (4.6 billion to 541 million years ago).

Law of Superposition

The law of superposition states that in an undisturbed collection of sediment or rock; the layer on the top is the youngest layer, while that at the bottom is the oldest. For instance, if in a given sample, a layer of limestone is present on top of a layer of shale stone, then the limestone layer is the younger of the two.

From this definition, it is clear that the law of superposition is concerned with the different layers of the Earth, and their respective ages. It helps in the determination of the age of a particular layer of earth, relative to the ages of the layers that are above and below it.

The law of superposition is one of the most fundamental principles of stratigraphy - the branch of archeology which studies the geological layers of the Earth and soil. This law has a proven record of being able to accurately determine the relative ages of the different layers of soil.

The formation of the different layers of soil is a result of the erosion of rocks over millions of years. Natural erosive agents such as wind, flowing water, etc., erode rocks, breaking off small particles from their surfaces. These particles, along with the minerals in them, get broken down into fine soil, which is then deposited on distant plains and other large land masses.

As this process continues, many years later, several layers of soil are formed one over the other on the site of deposition, just like the cards in a deck. The card at the bottom of the deck is always placed first while, and as the stack gets built, the one on the top is placed last. Similarly, the topmost layer of soil is the most recent one, while the bottommost is the oldest among the different layers of soil. This forms the basis of the working on the law of superposition.

The law implies that each layer of soil is younger than the one below it, and conversely, each layer of soil is older than the one that is above it. The law is also applicable to volcanic formations, wherein, molten magma erupts and spreads over older flows, solidifying into a new layer. In such formations too, each new layer formed is younger than the one below it.

Advantages of Superposition

The law of superposition provides us with a window into time, through which we can not only watch the events of the ages past, but also establish a chronologically correct sequence of their occurrences.

For example, if a layer of soil showing evidences of a flood is found to be deposited above another one that provides evidences of a famine, we can establish with certainty that a period of famine was followed by the occurrence of a flood.

Thus, the law of superposition helps in relative dating of various layers of soil, which can be used to establish a chronology of the occurrence of events at the different points of time, distinguished by the different deposition times of each layer.

The law also plays an important role in the field of archeology. It is considered the best method used by archaeologists for determining the relatives ages of various archaeological materials and artifacts that are dug out from the earth.

Limitations of Superposition

Though the law of superposition is an accurate way of determining the ages of the layers of earth relative to one another, it has a few inherent flaws. The definition of the law itself indicates that it is valid only for the dating of an 'undisturbed' sample of soil or rock. Thus, application of the law to earthen layers disturbed by natural occurrences, such as landslides, upheavals, cave-ins, or the creation of thrust faults which can cause old strata to overlay the younger one, can give misleading results.

Large-scale excavations by humans in the past can also disturb the layers of soil and rock to such an extent, that it can confound the outcome of the law's application. Similarly, various environmental reasons could have caused a period of time to pass without any sedimentary deposition. This could adversely affect the accurate establishment of a historical timeline.

Principle of Lateral Continuity

The principle of lateral continuity states that layers of sediment initially extend laterally in all directions; in other words, they are laterally continuous. As a result, rocks that are otherwise similar, but are now separated by a valley or other erosional feature, can be assumed to be originally continuous.

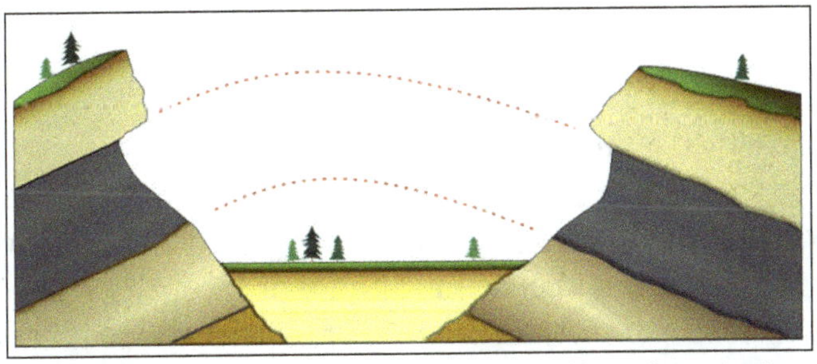

Schematic representation of the principle of lateral continuity.

Layers of sediment do not extend indefinitely; rather, the limits can be recognized and are controlled by the amount and type of sediment available and the size and shape of the sedimentary basin. As long as sediment is transported to an area, it will eventually be deposited. However, as the amount of material lessens away from the source, the layer of that material will become thinner.

Law of Included Fragments

The law of included fragments is a method of relative dating in geology. Essentially, this law states that clasts in a rock are older than the rock itself. One example of this is a xenolith, which is a fragment of country rock that fell into passing magma as a result of stoping. Another example is a derived fossil, which is a fossil that has been eroded from an older bedand redeposited into a younger one.

This is a restatement of Charles Lyell's original principle of inclusions and components from his 1830 to 1833 multi-volume *Principles of Geology*, which states that, with sedimentary rocks, if inclusions (or clasts) are found in a formation, then the inclusions must be older than the formation that contains them. For example, in sedimentary rocks, it is common for gravel from an older formation to be ripped up and included in a newer layer. A similar situation with igneous rocks occurs when xenoliths are found. These foreign bodies are picked up as magma or lava flows, and are incorporated, later to cool in the matrix. As a result, xenoliths are older than the rock which contains them.

Principle of Faunal Succession

The principle of faunal succession, also known as the law of faunal succession, is based on the observation that sedimentary rock strata contain fossilized flora and fauna, and that these fossils succeed each other vertically in a specific, reliable order that can be identified over wide horizontal distances. A fossilized Neanderthal bone will never be found in the same stratum as a fossilized Megalosaurus, for example, because neanderthals and megalosauruses lived during different geological periods, separated by many millions of years. This allows for strata to be identified and dated by the fossils found within.

This principle, which received its name from the English geologist William Smith, is of great importance in determining the relative age of rocks and strata. The fossil content of rocks together with the law of superposition helps to determine the time sequence in which sedimentary rocks were laid down.

Evolution explains the observed faunal and floral succession preserved in rocks. Faunal succession was documented by Smith in England during the first decade of the 19th century, and concurrently in France by Cuvier (with the assistance of the mineralogist Alexandre Brongniart). Archaic biological features and organisms are succeeded in the fossil record by more modern versions. For instance, paleontologists investigating the evolution of birds predicted that feathers would first be seen in primitive forms on flightless predecessor organisms such as feathered dinosaurs. This is precisely what has been discovered in the fossil record: simple feathers, incapable of supporting flight, are succeeded by increasingly large and complex feathers.

In practice, the most useful diagnostic species are those with the fastest rate of species turnover and the widest distribution; their study is termed biostratigraphy, the science of dating rocks by using the fossils contained within them. In Cenozoic strata, fossilized tests of foraminifera are often used to determine faunal succession on a refined scale, each biostratigraphic unit (biozone) being a geological stratum that is defined on the basis of its characteristic fossil taxa. An outline microfaunal zonal scheme based on both foraminifera and ostracoda was compiled by M. B. Hart. Simply, the earlier fossil life forms are simpler than more recent forms, and more recent forms are most similar to existing forms (principle of faunal succession).

References

- Stratigraphy: uwsp.edu, Retrieved 16 April, 2019
- Principle-of-original-horizontality: geologyin.com, Retrieved 16 June, 2019
- Uniformitarianism: britannica.com, Retrieved 27 April, 2019
- What-does-law-of-superposition-mean: sciencestruck.com, Retrieved 16 June, 2019
- The-principle-of-lateral-continuity: geologyin.com, Retrieved 09 January, 2019
- Principle-of-faunal-succession: geology.fandom.com, Retrieved 26 March, 2019

Branches of Geology 4

- Startigraphy
- Structural Geology
- Petrology
- Ethnogeology
- Historical Geology
- Geochronology

Geology is a vast discipline which can be divided into numerous branches. A few of these are stratigraphy, structural geology, petrology, ethnogeology, historical geology and geochronology. This chapter has been carefully written to provide an easy understanding of these branches of geology.

Startigraphy

Stratigraphy is that subarea of geology that treats the description, correlation and interpretation of stratified Earth materials. Typically, geologists consider stratified Earth materials as layers of sediment or sedimentary rock. This definition, however, clearly encompasses other materials such as volcanic lava, ash flows, ash-fall layers, meteoritic impact ejecta layers, and soils. In fact, using this definition, any material that obeys the law of superposition during its formation could be placed in the domain of stratigraphy. Generally, internal layers within Earth (crust, mantle, and core) are not considered the type of layers studied by stratigraphers because they formed by Earth's internal differentiation processes.

Some geologists give a broader definition to the term stratigraphy. Planetary geologists sometimes view stratigraphy as if it were the study of the sequence of events on a planet or moon's surface. In addition, stratigraphy has been broadly used by some geologists who study mountain building and plate tectonics to mean the study of order of emplacement of rock units of various types, including igneous and metamorphic rocks, to which the law of superposition does not apply. In some cases, stratigraphy is used to define the study of geologic history of an area or country, but it is more correct to say that stratigraphy is the practical foundation for historical geology. In this topic the concept of stratigraphy expressed in the first paragraph is viewed as best and most correct.

Stratigraphy had its origins in the Renaissance writings of Nicholas Steno, who was the first to write lucidly about sedimentary strata. He observed strata exposed in the Arno River valley of Italy, and noted three axiomatic ideas, which became known as the first three "laws" of stratigraphy. These laws are known today as superposition, lateral continuity, and original horizontality. Superposition holds that layers are deposited so that the older layer is on the bottom. Unless strata are disturbed, this is always true. Lateral continuity holds that sedimentary layers extend laterally until they become so thin that they end at a "feather edge," abut against an obstruction, or grade laterally into other layers. Original horizontality holds that sedimentary layers are originally formed horizontally and remain so unless deformed by subsequent processes.

Steno's writings were full of common sense. In super-position, he noted the most important criterion for relative age dating. In lateral continuity, he wrote about how correlation of sedimentary layers would be possible. In original horizontality, he noted the criterion necessary for any sort of analysis of later deformation, that is, the original state of a sedimentary layer can be assumed to be horizontal.

As insightful as Steno's writings were, there is no strong evidence that they were influential beyond the Renaissance era in which he lived. Later on, during the Enlightenment, naturalists like James Hutton, John Playfair and Charles Lyell apparently independently "re-discovered" the importance of these common-sense concepts and used them in their influential writings about geology and stratigraphy. Hutton, Playfair, Lyell, and others of their time wrote books and papers, which established the foundations of modern thought about stratigraphy. Their most important contributions included promoting the concepts of actualism (understanding the past by studying modern processes) and demonstrating such key concepts as stratigraphic correlation, predictable fossil succession, and the great antiquity of Earth.

The advancement of these key concepts were given a great boost by the pioneering work of the English field engineer William Smith, who compiled and published the first large-scale geologic map employing modern concepts of stratigraphic correlation and fossil succession. Smith's success inspired others to this kind of work, and was particularly important in influencing the Geological Society of London to embark upon its "stratigraphical enterprise" of research in the United Kingdom. The Society and the British Geological Survey were important promoters of early stratigraphic studies and venues for presentation of early research. Based upon these efforts, it is fair to assert that modern stratigraphy was born in the United Kingdom during this period.

In the nineteenth century, major efforts were made by British stratigraphers and their colleagues on the European continent to develop a unified stratigraphic succession (or "geological column") for rocks in their areas. Cambridge Professor Adam Sedgwick and Scottish naturalist Roderick Murchison became quite famous as the preeminent "system builders" of their time. Sedgwick studied and named the Cambrian System himself and with Murchison, the Devonian. Murchison studied and named the Silurian and Permian Systems by himself. There were others who did the same during the nineteenth century, thus establishing the basis of our modern geological time scale (which has periods of the same names as those given to "systems" of rock during an era when exact ages of rock strata were unknown). This was the birth of modern chronostratigraphy, which emphasizes subdivision of geological time by studying Earth's stratigraphic record.

A Swiss geologist, Amanz Gressley, studied Jurassic strata in Europe in hopes of understanding what happens to sedimentary layers where they grade into other layers. He recognized that lateral continuity of layers revealed many changes, which reflected different ancient environments. To this concept, he gave the name facies, meaning an aspect of a sedimentary formation. A German stratigrapher, Johannes Walther, took up Gressley's ideas in his own work and became more widely known than Gressley for work with sedimentary facies. To Walther, the facies represented primary characteristics of the rock that would help him understand how and where the rock formed. He used what he called the ontological method in facies stratigraphy, which he described whimsically as "from being, we explain becoming." This was a direct application of actualism, advocated earlier by Hutton and others, but now applied in a time of enhanced understanding of the natural world. Walter was the first naturalist to spend large amounts of time in the field studying modern environments in order to better interpret the past. His two-volume work, Modern Lithogenesis, was a watershed for modern research with sedimentary facies. Accordingly, Walter is regarded as the founder of modern facies stratigraphy. Although his work was not accepted well in the United States for many years (due, in part, to anti-German feelings during the early twentieth century), it later was studied extensively for its rich descriptions of modern sedimentary environments and ancient sedimentary facies. In the latter part of the twentieth century, facies stratigraphy became much more than an academic exercise when it was realized that such knowledge could help predict the occurrence of petroleum and certain ore minerals—and facilitate more productive extraction of these materials—in host sedimentary rocks.

At the outset of the twentieth century, Austrian stratigrapher Eduard Suess became the first advocate of global changes of sea level and how those changes might relate to global stratigraphy. This concept, called eustatsy, holds that global sea level rises and falls during geological history lead to the great marine transgressions and regressions noted in many sedimentary strata from locales around the world. Suess called upon subsidence of the sea floor and displacement of seawater by sediment as reasons for this global effect (today we know that gain and loss of polar ice is another contributor to sea-level change). His work stimulated much research, and strongly influenced the well-known American geologist T.C. Chamberlain, who perpetuated these ideas through his many well-known papers on the subject. These ideas were important in the development of a modern concept in stratigraphy called sequence stratigraphy.

Sequence stratigraphy, which holds that large bodies of sedimentary strata are bounded by interregional unconformities, formed as a result of global eustatsy. In the early 1960s, sequence stratigraphy was put forth by the American stratigrapher L.L. Sloss in a series of widely read papers. During the 1970s, Sloss's student, Peter Vail, formerly with Exxon Corporation (now Exxon-Mobil Corporation), further developed these concepts while studying seismic profiles of stratigraphy from the world's continental shelves. Vail's paper's established sequence stratigraphy as one of the main subdivisions of modern stratigraphy. To recognize their contributions, sequence stratigraphy is often referred to as Sloss-Vail sequence stratigraphy in their honor.

Vail's work spawned a huge effort to produce a highly detailed, eustatic sea-level cycle chart of Earth's history based upon the vast data collection at Exxon. His work was published in 1987 in the prestigious journal Science. Sequence stratigraphy and global sea-level cycle charts are concepts used today major petroleum-company exploration laboratories all over the world.

Today, facies stratigraphy and sequence stratigraphy are not the only types of stratigraphy practiced by geologists. Modern stratigraphy includes: lithostratigraphy (naming of formations for purposes of geological mapping); biostratigraphy (correlating rock layers based upon fossil content); chronostratigraphy (correlating rock layers based upon their similar ages); magnetostratigraphy (study and correlation of rock layers based upon their inherent magnetic character); soil stratigraphy (study and mapping of soil layers, modern and ancient); and event stratigraphy (study and correlation of catastrophic events in geological history). The latter may include global or regional layers formed by asteroid or comet impacts, major volcanic events, global climate or ocean-chemistry changes, and effects of slight changes in Earth's orbital parameters (e.g., Milankovitch cycles).

Because layered Earth materials possess so much information about Earth's past, including the entire fossil record— and a sedimentary record quite sensitive to atmospheric, climatic, and oceanic changes of the past—stratigraphy is the one subarea of geology entirely focused upon retrieving and understanding that record.

Structural Geology

Structural geology is the study of the three-dimensional distribution of rock units with respect to their deformational histories. The primary goal of structural geology is to use measurements of present-day rock geometries to uncover information about the history of deformation (strain) in the rocks, and ultimately, to understand the stress field that resulted in the observed strain and geometries. This understanding of the dynamics of the stress field can be linked to important events in the geologic past; a common goal is to understand the structural evolution of a particular area with respect to regionally widespread patterns of rock deformation (e.g., mountain building, rifting) due to plate tectonics.

Use and Importance

The study of geologic structures has been of prime importance in economic geology, both petroleum geology and mining geology. Folded and faulted rock strata commonly form traps that accumulate and concentrate fluids such as petroleum and natural gas. Similarly, faulted and structurally complex areas are notable as permeable zones for hydrothermal fluids, resulting in concentrated areas of base and precious metal ore deposits. Veins of minerals containing various metals commonly occupy faults and fractures in structurally complex areas. These structurally fractured and faulted zones often occur in association with intrusive igneous rocks. They often also occur around geologic reef complexes and collapse features such as ancient sinkholes. Deposits of gold, silver, copper, lead, zinc, and other metals, are commonly located in structurally complex areas.

Structural geology is a critical part of engineering geology, which is concerned with the physical and mechanical properties of natural rocks. Structural fabrics and defects such as faults, folds, foliations and joints are internal weaknesses of rocks which may affect the stability of human engineered structures such as dams, road cuts, open pit mines and underground mines or road tunnels.

Geotechnical risk, including earthquake risk can only be investigated by inspecting a combination of structural geology and geomorphology. In addition, areas of karst landscapes which reside atop underground caverns, potential sinkholes, or other collapse features are of particular importance for these scientists. In addition, areas of steep slopes are potential collapse or landslide hazards.

Environmental geologists and hydrogeologists need to apply the tenets of structural geology to understand how geologic sites impact (or are impacted by) groundwater flow and penetration. For instance, a hydrogeologist may need to determine if seepage of toxic substances from waste dumps is occurring in a residential area or if salty water is seeping into an aquifer.

Plate tectonics is a theory developed during the 1960s which describes the movement of continents by way of the separation and collision of crustal plates. It is in a sense structural geology on a planet scale, and is used throughout structural geology as a framework to analyze and understand global, regional, and local scale features.

Methods

Structural geologists use a variety of methods to (first) measure rock geometries, (second) reconstruct their deformational histories, and (third) estimate the stress field that resulted in that deformation.

Geometries

Primary data sets for structural geology are collected in the field. Structural geologists measure a variety of planar features (bedding planes, foliation planes, fold axial planes, fault planes, and joints), and linear features (stretching lineations, in which minerals are ductily extended; fold axes; and intersection lineations, the trace of a planar feature on another planar surface).

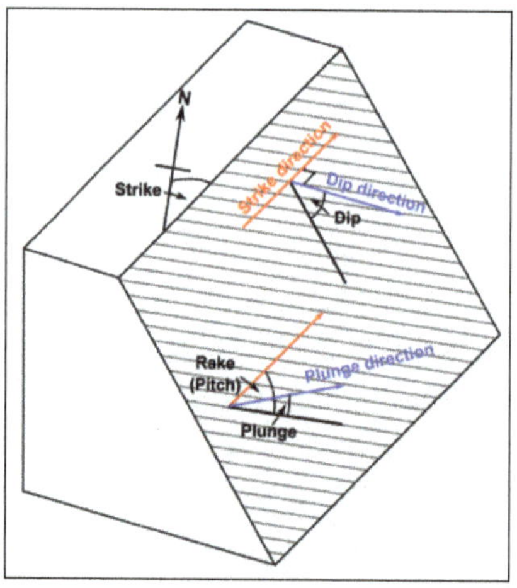

Illustration of measurement conventions for planar and linear structures.

Measurement Conventions

The inclination of a planar structure in geology is measured by *strike and dip*. The strike is the line of intersection between the planar feature and a horizontal plane, taken according to the right hand convention, and the dip is the magnitude of the inclination, below horizontal, at right angles to strike. For example; striking 25 degrees East of North, dipping 45 degrees Southeast, recorded as N25E,45SE. Alternatively, dip and dip direction may be used as this is absolute. Dip direction is measured in 360 degrees, generally clockwise from North. For example, a dip of 45 degrees towards 115 degrees azimuth, recorded as 45/115.

The term *hade* is occasionally used and is the deviation of a plane from vertical i.e. (90°-dip).

Fold axis plunge is measured in dip and dip direction (strictly, plunge and azimuth of plunge). The orientation of a fold axial plane is measured in strike and dip or dip and dip direction.

Lineations are measured in terms of dip and dip direction, if possible. Often lineations occur expressed on a planar surface and can be difficult to measure directly. In this case, the lineation may be measured from the horizontal as a *rake* or *pitch* upon the surface.

Rake is measured by placing a protractor flat on the planar surface, with the flat edge horizontal and measuring the angle of the lineation clockwise from horizontal. The orientation of the lineation can then be calculated from the rake and strike-dip information of the plane it was measured from, using a stereographic projection.

If a fault has lineations formed by movement on the plane, e.g. slickensides, this is recorded as a lineation, with a rake, and annotated as to the indication of throw on the fault.

Generally it is easier to record strike and dip information of planar structures in dip/dip direction format as this will match all the other structural information you may be recording about folds, lineations, etc., although there is an advantage to using different formats that discriminate between planar and linear data.

Plane, Fabric, Fold and Deformation Conventions

The convention for analysing structural geology is to identify the planar structures, often called *planar fabrics* because this implies a textural formation, the linear structures and, from analysis of these, unravel deformations.

Planar structures are named according to their order of formation, with original sedimentary layering the lowest at So. Often it is impossible to identify So in highly deformed rocks, so numbering may be started at an arbitrary number or given a letter (S_A, for instance). In cases where there is a bedding-plane foliation caused by burial metamorphism or diagenesis this may be enumerated as Soa.

If there are folds, these are numbered as F_1, F_2, etc. Generally the axial plane foliation or cleavage of a fold is created during folding, and the number convention should match. For example, an F_2 fold should have an S_2 axial foliation.

Deformations are numbered according to their order of formation with the letter D denoting a deformation event. For example, D_1, D_2, D_3. Folds and foliations, because they are formed by deformation events, should correlate with these events. For example, an F_2 fold, with an S_2axial plane foliation would be the result of a D_2 deformation.

Metamorphic events may span multiple deformations. Sometimes it is useful to identify them similarly to the structural features for which they are responsible, e.g.; M_2. This may be possible by observing porphyroblast formation in cleavages of known deformation age, by identifying metamorphic mineral assemblages created by different events, or via geochronology.

Intersection lineations in rocks, as they are the product of the intersection of two planar structures, are named according to the two planar structures from which they are formed. For instance, the intersection lineation of a S_1 cleavage and bedding is the $L_{1\text{-}0}$ intersection lineation (also known as the cleavage-bedding lineation).

Stretching lineations may be difficult to quantify, especially in highly stretched ductile rocks where minimal foliation information is preserved. Where possible, when correlated with deformations (as few are formed in folds, and many are not strictly associated with planar foliations), they may be identified similar to planar surfaces and folds, e.g.; L_1, L_2. For convenience some geologists prefer to annotate them with a subscript S, for example L_{s1} to differentiate them from intersection lineations, though this is generally redundant.

Stereographic Projections

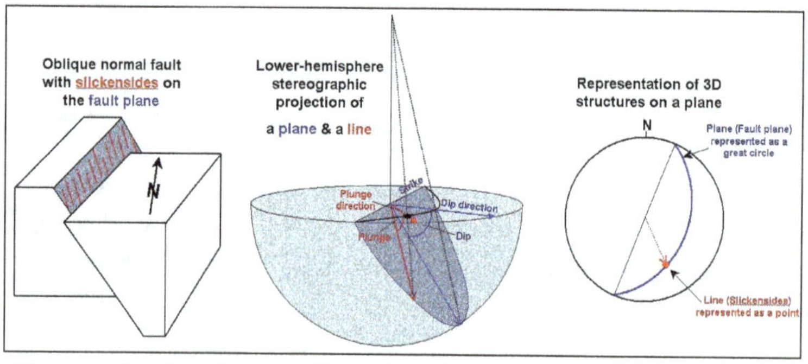

Diagram showing the use of lower hemisphere stereographic projection in structural geology using an example of a fault plane and a slickenside lineationobserved within the fault plane.

Stereographic projection is a method for analyzing the nature and orientation of deformation stresses, lithological units and penetrative fabrics wherein linear and planar features (structural strike and dip readings, typically taken using a compass clinometer) passing through an imagined sphere are plotted on a two-dimensional grid projection, facilitating more holistic analysis of a set of measurements.

Rock Macro-structures

On a large scale, structural geology is the study of the three-dimensional interaction and relationships of stratigraphic units within terranes of rock or geological regions.

This branch of structural geology deals mainly with the orientation, deformation and relationships of stratigraphy (bedding), which may have been faulted, folded or given a foliation by some tectonic event. This is mainly a geometric science, from which *cross sections* and three-dimensional *block models* of rocks, regions, terranes and parts of the Earth's crust can be generated.

Study of regional structure is important in understanding orogeny, plate tectonics and more specifically in the oil, gas and mineral exploration industries as structures such as faults, folds and unconformities are primary controls on ore mineralisation and oil traps.

Modern regional structure is being investigated using seismic tomography and seismicreflection in three dimensions, providing unrivaled images of the Earth's interior, its faults and the deep crust. Geophysics such as gravity and airborne magnetics can provide information on the nature of rocks imaged to be in the deep crust.

Rock Microstructures

Rock microstructure or *texture* of rocks is studied by structural geologists on a small scale to provide detailed information mainly about metamorphic rocks and some features of sedimentary rocks, most often if they have been folded. Textural study involves measurement and characterisation of foliations, crenulations, metamorphic minerals, and timing relationships between these structural features and mineralogical features. Usually this involves collection of hand specimens, which may be cut to provide petrographicthin sections which are analysed under a petrographic microscope. Microstructural analysis finds application also in multi-scale statistical analysis, aimed to analyze some rock features showing scale invariance.

Kinematics

Geologists use rock geometry measurements to understand the history of strain in rocks. Strain can take the form of brittle faulting and ductile folding and shearing. Brittle deformation takes place in the shallow crust, and ductile deformation takes place in the deeper crust, where temperatures and pressures are higher.

Stress Fields

By understanding the constitutive relationships between stress and strain in rocks, geologists can translate the observed patterns of rock deformation into a stress field during the geologic past. The following list of features are typically used to determine stress fields from deformational structures.

- In perfectly brittle rocks, faulting occurs at 30° to the greatest compressional stress. (Byerlee's Law)
- The greatest compressive stress is normal to fold axial planes.

Characterization of the Mechanical Properties of Rock

The mechanical properties of rock play a vital role in the structures that form during deformation deep below the earth's crust. The conditions in which a rock is present will result in different structures that geologists observe above ground in the field. The field of structural geology tries to relate the formations that humans see to the changes the rock went through to get to that final structure. Knowing the conditions of deformation that lead to such structures can illuminate the history of the deformation of the rock.

Temperature and pressure play a huge role in the deformation of rock. At the conditions under the earth's crust of extreme high temperature and pressure, rocks are ductile. They can bend, fold or break. Other vital conditions that contribute to the formation of structure of rock under the earth are the stress and strain fields.

Stress-strain Curve

Stress is a pressure, defined as a directional force over area. When a rock is subjected to stresses, it changes shape. When the stress is released, the rock may or may not return to its original shape. That change in shape is quantified by strain, the change in length over the original length of the material in one dimension. Stress induces strain which ultimately results in a changed structure.

Elastic deformation refers to a reversible deformation. In other words, when stress on the rock is released, the rock returns to its original shape. Reversible, linear, elasticity involves the stretching, compressing, or distortion of atomic bonds. Because there is no breaking ofbonds, the material springs back when the force is released. This type of deformation is modeled using a linear relationship between stress and strain, i.e. a Hookean relationship.

$$\epsilon = \frac{\sigma}{\mathrm{E}}$$

Where, σ denotes stress, ϵ denotes strain, and E is the elastic modulus, which is material dependent. The elastic modulus is, in effect, a measure of the strength of atomic bonds.

Plastic deformation refers to non-reversible deformation. The relationship between stress and strain for permanent deformation is nonlinear. Stress has caused permanent change of shape in the material by involving the breaking of bonds.

One mechanism of plastic deformation is the movement of dislocations by an applied stress. Because rocks are essentially aggregates of minerals, we can think of them as poly-crystalline materials. Dislocations are a type of crystallographic defect which consists of an extra or missing half plane of atoms in the periodic array of atoms that make up a crystal lattice. Dislocations are present in all real crystallographic materials.

Hardness

Hardness is difficult to quantify. It is a measure of resistance to deformation, specifically permanent deformation. There is precedent for hardness as a surface quality, a measure of the abrasiveness or surface-scratching resistance of a material. If the material being tested, however, is uniform in composition and structure, then the surface of the material is only a few atomic layers thick, and measurements are of the bulk material. Thus, simple surface measurements yield information about the bulk properties. Ways to measure hardness include:

- Mohs Scale
- Dorry abrasion test
- Deval abrasion test
- Indentation hardness

Indentation hardness is used often in metallurgy and materials science and can be thought of as resistance to penetration by an indenter.

Toughness

Toughness can be described best by a material's resistance to cracking. During plastic deformation, a material absorbs energy until fracture occurs. The area under the stress-strain curve is the work required to fracture the material. The toughness modulus is defined as,

$$M_t = \frac{2}{3}\sigma_{UTS}\,\epsilon_f$$

Where, σ_{UTS} is the ultimate tensile strength, and ϵ_f is the strain at failure. The modulus is the maximum amount of energy per unit volume a material can absorb without fracturing. From the equation for modulus, for large toughness, high strength and high ductility are needed. These two properties are usually mutually exclusive. Brittle materials have low toughness because low plastic deformation decreases the strain (low ductility). Ways to measure toughness include:

- Page impact machine
- Charpy Impact Test

Resilience

Resilience is a measure of the elastic energy absorbed of a material under stress. In other words, the external work performed on a material during deformation. The area under the elastic portion of the stress-strain curve is the strain energy absorbed per unit volume. The resilience modulus is defined as,

$$M_R = \frac{(\sigma_y)^2}{2E}$$

where, σ_y is the yield strength of the material and E is the elastic modulus of the material. To increase resilience, one needs increased elastic yield strength and decreased modulus of elasticity.

Petrology

Petrology refers to the scientific study of rocks and the conditions which influence their formation. Petrology is a branch of geology that focuses on the chemical analysis in various fields such as petrography and mineralogy. By incorporating various principles of geophysics and geochemistry, modern petrologists can establish the origins of rocks and their chemical characteristics. There are three main branches of petrology resultant from the main rock types: sedimentary, igneous, and metamorphic.

Importance of Petrology

Petrology plays an important role in ascertaining the physical and chemical composition of rocks and the different conditions that influence their formation. Modern petrologists rely on knowledge in mineralogy to help in mapping and sampling of rocks. Since most rocks constitute various minerals, it becomes easier to study and understand them with background knowledge in mineralogy.

Petrological research is also crucial in helping us understand the nature of the earth's crust and mantle. Over many years, the earth's tectonic processes have shaped the nature of rocks and the topography of the earth's surface. These geological processes are vital in determining the suitability of certain areas for agriculture, industrial or commercial use.

Aside from helping us understand the history of our planet's formation, petrology applies in many areas that directly affect our livelihood. The field of petrology helps us to understand the best raw materials to be used in industries for the manufacture of goods. Additionally, petrology promotes best practices that foster sustenance and technological advancement.

Branches of Petrology

There are three main branches of petrology, namely: igneous, sedimentary, and metamorphic.

Igneous Petrology

Igneous petrology is a branch that specializes in the scientific study of igneous rocks, their chemical composition and texture. Igneous rocks are rocks formed when magma or molten rock is crystallized to form granite or basalt. Igneous rocks may be formed through crystallization or may involve a different process that cools or solidifies the molten rock. Igneous rocks that are formed beneath the earth's surface form 'intrusive' rocks, also known as plutonic rocks. Rocks found on the earth's surface produce extrusive rocks. About 95 percent of the earth's crust is made up of igneous rocks. However, these rocks do not readily appear on the surface because they are covered by a thin extensive layer of sedimentary and metamorphic rocks.

Sedimentary Petrology

This category focuses on the study of sedimentary rocks with an emphasis on texture and composition. Sedimentary rocks include shale, limestone, and sandstone. They are formed when particles derived from other rocks are bound together to form a matrix. These pieces of rock particles together with deposits of chemical and biological properties coalesce to produce a fine rock material.

Metamorphic Petrology

Metamorphic petrology involves the study of the composition and texture of metamorphic rocks. Such rocks include marble, gneiss, slate, and schist. Initially, sedimentary or igneous rocks undergo mineralogical, textural and chemical changes because of extreme pressure or temperature, and in some cases due to both. Due to high pressure and temperatures, the igneous or sedimentary rocks are transformed into metamorphic rocks.

Experimental Petrology

Experimental petrology involves the study of natural or synthetic materials when subjected to conditions of high pressure and high temperature. This helps to establish the phase relations and geochemical properties of the materials. The work of experimental petrology has provided a firm foundation for understanding the processes involved in the formation of metamorphic rocks. Rocks on the lower crust and upper mantle of the earth can be easily studied and analyzed as a result of experimental petrology.

Practical Application of Petrology

Petrology and its application are important for several reasons:

- The study of rocks helps us to understand the composition of the earth's crust. The various minerals and chemical properties of the mantle are important to petrologist. Various professions such as civil engineering rely heavily on the findings of petrologists.
- By studying rocks, historical formation of the earth's surface can be established. Methods such as radiometric dating can be used to determine the time periods in which certain rock types were formed. This helps in creating a comprehensive series of geological events.
- Scientists can now understand the tectonic processes that led to the formation of the earth's surface. This is because specific tectonic environments have distinct characteristics from others. Scientists can, therefore, reconstitute these processes.
- Rocks are natural sources for important minerals and ores which are used as raw materials by industries. Petrologists help in the discovery of new minerals hidden below the earth's surface. The supply of such minerals is also vital to ensuring sustenance and commercial viability.
- The study of the earth's rocks helps in technological advancement.

Ethnogeology

Ethnogeology is the study of how geological features were understood by ancient peoples around the globe from a "place-based" perspective, in specific reference to traditional knowledge and to the stories and ideas about the Earth that were passed down through traditions and the wisdom of elders. The focus in past research tended to be on the unique ideas and knowledge of minorities and distinct cultural groups, and how this relates to universal and cross cultural knowledge discovered by humanity as a whole. Knowledge claims that are based more on universal discoveries and natural science can be found in the subjects of philosophy, chemistry, physics, biology, mathematics, and geology. The term "ethnogeology" first enters the geological literature through the work of John Murray of the University of Manitoba in Winnipeg, Canada in the mid-1990s by virtue of his studies of the Northern Cree First Nation geological worldviews in the Province of Manitoba in Canada.At that time, a cadre of geologists and geoscience educators - particularly Dr. Steven Semken of Navajo Community College in Shiprock, New Mexico and colleagues - were examining the interesting connections among Native American traditional knowledge, geoscience concepts, and the unique vision of planet Earth's history as articulated by indigenous ethnogeological wisdom.

One global example of geological knowledge over thousands of years is the making of rock tools and arrowheads. Worldwide, humans have shaped specific kinds of rock into various tools. These artifacts are evidence that there was some basic knowledge of which rocks were hard enough and easy enough to chip into various useful forms. Ancient humans discovered certain rock formations that provided them with materials to work with. Flint, obsidian, chert, copper and quartz were some examples of useful rocks or minerals for thousands of years.

In addition, humans have always used natural, high, and dry ridges for land transportation. Natural water ways were used for boats, travel, exploration, and trade. Various types of natural elements were used around the Earth for centuries. Gold and silver are prized for jewelry and trade. Good soil is needed for agriculture. Mountains and rivers provided natural boundaries, defence, and borders for various political groups.

Historical Geology

Historical geology or paleogeology is a discipline that uses the principles and techniques of geology to reconstruct and understand the geological history of Earth. It focuses on geologic processes that change the Earth's surface and subsurface; and the use of stratigraphy, structural geology and paleontology to tell the sequence of these events. It also focuses on the evolution of plants and animals during different time periods in the geological timescale. The discovery of radioactivity and the development of several radiometric dating techniques in the first half of the 20th century provided a means of deriving absolute versus relative ages of geologic history.

Economic geology, the search for and extraction of fuel and raw materials, is heavily dependent on an understanding of the geological history of an area. Environmental geology, including most importantly the geologic hazards of earthquakes and volcanism, must also include a detailed knowledge of geologic history.

Nicolaus Steno, also known as Niels Stensen, was the first to observe and propose some of the basic concepts of historical geology, called the "father of geology". One of these concepts was that fossils originally came from living organisms. The other, more famous, observations are often grouped together to form the laws of stratigraphy.

James Hutton and Charles Lyell also contributed to early understanding of the Earth's historywith their observations at Edinburgh in Scotland concerning angular unconformity in a rock face and it was in fact Lyell that influenced Charles Darwin greatly in his theory of evolution by speculating that *the present is the key to the past.* Hutton first proposed the theory of uniformitarianism, which is now a basic principle in all branches of geology. Hutton also supported the idea that the Earth was very old as opposed to the prevailing concept of the time which said the Earth had only been around a few millennia. Uniformitarianism describes an Earth created by the same natural phenomena that are at work today.

The prevailing concept of the 18th century in the West was that of a very small age of the Earth, dominated by catastrophic events. This view was strongly supported by adherents of Abrahamic religions based on a largely literal interpretation of their religious scriptural passages. The concept of uniformitarianism met with considerable resistance and the catastrophism vs. gradualism debate of the 19th century resulted. A variety of discoveries in the 20th century provided ample evidence that Earth history is a product of both gradual incremental processes and sudden cataclysmic events. Violent events such as meteorite impacts and large volcanic explosions do shape the Earth's surface along with gradual processes such as weathering, erosion and deposition much as they have throughout Earth history. The present is the key to the past - includes catastrophic as well as gradual processes.

Geochronology

Geochronology is the science of determining the age of rocks, fossils, and sediments using signatures inherent in the rocks themselves. Absolute geochronology can be accomplished through radioactive isotopes, whereas relative geochronology is provided by tools such as palaeomagnetism and stable isotope ratios. By combining multiple geochronological (and biostratigraphic) indicators the precision of the recovered age can be improved.

An artistic depiction of the major events in the history of Earth.

Geochronology is different in application from biostratigraphy, which is the science of assigning sedimentary rocks to a known geological period via describing, cataloging and comparing fossil floral and faunal assemblages. Biostratigraphy does not *directly* provide an absolute age determination of a rock, but merely places it within an *interval*of time at which that fossil assemblage is known to have coexisted. Both disciplines work together hand in hand, however, to the point where they share the same system of naming rock layers and the time spans utilized to classify layers within a stratum.

The science of geochronology is the prime tool used in the discipline of chronostratigraphy, which attempts to derive absolute age dates for all fossil assemblages and determine the geologic history of the Earth and extraterrestrial bodies.

Dating Methods

Cosmogenic Nuclide Geochronology

A series of related techniques for determining the age at which a geomorphic surface was created (exposure dating), or at which formerly surficial materials were buried (burial dating). Exposure dating uses the concentration of exotic nuclides (e.g. ^{10}Be, ^{26}Al, ^{36}Cl) produced by cosmic rays interacting with Earth materials as a proxy for the age at which a surface, such as an alluvial fan, was created. Burial dating uses the differential radioactive decay of 2 cosmogenic elements as a proxy for the age at which a sediment was screened by burial from further cosmic rays exposure.

Luminescence Dating

Luminescence dating techniques observe 'light' emitted from materials such as quartz, diamond, feldspar, and calcite. Many types of luminescence techniques are utilized in geology, including optically stimulated luminescence (OSL), cathodoluminescence (CL), and thermoluminescence (TL). Thermoluminescence and optically stimulated luminescence are used in archaeology to date ‹fired' objects such as pottery or cooking stones and can be used to observe sand migration.

Incremental Dating

Incremental dating techniques allow the construction of year-by-year annual chronologies, which can be fixed (*i.e.* linked to the present day and thus calendar or sidereal time) or floating.

- Dendrochronology
- Ice cores
- Lichenometry
- Varves

Paleomagnetic Dating

A sequence of paleomagnetic poles (usually called virtual geomagnetic poles), which are already well defined in age, constitutes an apparent polar wander path (APWP). Such a path is constructed for a large continental block. APWPs for different continents can be used as a reference for newly obtained poles for the rocks with unknown age. For paleomagnetic dating, it is suggested to use the APWP in order to date a pole obtained from rocks or sediments of unknown age by linking the paleopole to the nearest point on the APWP. Two methods of paleomagnetic dating have been suggested (1) Angular method and (2) Rotation method.First method is used for paleomagnetic dating of rocks inside of the same continental block. The second method is used for the folded areas where tectonic rotations are possible.

Magnetostratigraphy

Magnetostratigraphy determines age from the pattern of magnetic polarity zones in a series of bedded sedimentary and/or volcanic rocks by comparison to the magnetic polarity timescale. The polarity timescale has been previously determined by dating of seafloor magnetic anomalies, radiometrically dating volcanic rocks within magnetostratigraphic sections, and astronomically dating magnetostratigraphic sections.

Chemostratigraphy

Global trends in isotope compositions, particularly Carbon 13 and strontium isotopes, can be used to correlate strata.

Correlation of Marker Horizons

Tephra horizons in south-central Iceland. The thick and light-to-dark coloured layer at the height of the volcanologists hands is a marker horizon of rhyolitic-to-basaltic tephrafrom Hekla.

Marker horizons are stratigraphic units of the same age and of such distinctive composition and appearance, that despite their presence in different geographic sites, there is certainty about their age-equivalence. Fossil faunal and floral assemblages, both marine and terrestrial, make for distinctive marker horizons.Tephrochronology is a method for geochemical correlation of unknown volcanic ash (tephra) to geochemically fingerprinted, dated tephra. Tephrais also often used as a dating tool in archaeology, since the dates of some eruptions are well-established.

Geological Hierarchy of Chronological Periodization

Geochronology - From largest to smallest:

1. Supereon
2. Eon
3. Era
4. Period
5. Epoch
6. Age
7. Chron

Differences from Chronostratigraphy

It is important not to confuse geochronologic and chronostratigraphic units.Geochronological units are periods of time, thus it is correct to say that *Tyrannosaurus rex*lived during the Late

Cretaceous Epoch. Chronostratigraphic units are geological material, so it is also correct to say that fossils of the genus *Tyrannosaurus* have been found in the Upper Cretaceous Series. In the same way, it is entirely possible to go and visit an Upper Cretaceous Series deposit – such as the Hell Creek deposit where the *Tyrannosaurus* fossils were found – but it is naturally impossible to visit the Late Cretaceous Epoch as that is a period of time.

References

- Stratigraphy, earth-and-environment-geology-and-oceanography-geology-and-oceanography: encyclopedia.com, Retrieved 16 April, 2019
- What-is-petrology: worldatlas.com, Retrieved 14 August, 2019
- B.A. van der Pluijm and S. Marshak (2004). Earth Structure - An Introduction to Structural Geology and Tectonics (2nd ed.). New York: W. W. Norton. p. 656. ISBN 0-393-92467-X
- Faure, G., and Mensing, D. 2005. "Isotopes - Principles and applications". 3rd Edition. J. Wiley & Sons. ISBN 0-471-38437-2

Geomorphology and its Processes 5

- Geomorphology
- Biogeomorphology
- Aeolian processes
- Fluvial processes
- Erosion and tectonics
- Weathering
- Diagenesis
- Metamorphism
- Mass Wasting

The scientific study of the origin and evolution of topographic and bathymetric features is referred to as geomorphology. The processes related to this field include aeolian processes, fluvial processes, weathering, metamorphism, etc. This chapter closely examines these processes of geomorphology to provide an extensive understanding of the subject.

Geomorphology

Geomorphology is concerned with the shaping of landforms, through such processes as subsidence and uplift, and with the classification and study of such landforms as mountains, volcanoes, and islands.

Working of Geomorphology

Geomorphology is an area of geology concerned with the study of landforms, with the forces and processes that have shaped them, and with the description and classification of various physical features on Earth. The term, which comes from the Greek words geo, or "Earth," and morph, meaning "form," was coined in 1893 by the American geologist William Morris Davis, who is considered the father of geomorphology.

During Davis's time, geomorphology was concerned primarily with classifying different structures on Earth's surface, examples of which include mountains and islands, discussed later in this essay. This view of geomorphology as an essentially descriptive, past-oriented area of study closely aligned with historical geology prevailed throughout the late nineteenth and early twentieth centuries.

By the mid-twentieth century, however, the concept of geomorphology inherited from Davis had fallen into disfavor, to be replaced by a paradigm, or model, oriented toward physical rather than historical geology. (These two principal branches of geology are concerned, in the first instance, with Earth's past and the processes that shaped it and, in the second instance, with Earth's current physical features and the processes that continue to shape it).

Rethinking Geomorphology

As reconceived in the 1950s and thereafter, geomorphology became an increasingly exact science. As has been typical of many sciences in their infancy, early geomorphology focused on description rather than prediction and tended to approach its subject matter in a qualitative fashion. The term qualitative suggests a comparison between qualities that are not defined precisely, such as "fast" and "slow" or "warm" and "cold." On the other hand, a quantitative approach, as has been implemented for geomorphology from the mid-twentieth century onward, centers on a comparison between precise quantities—for instance, 10 lb. (4.5 kg) versus 100 lb. (45 kg) or 50 MPH (80.5 km/h) versus 120 MPH (193 km/h).

As part of its shift in focus, geomorphology began to treat Earth's physical features as systems made up of complex and ongoing interactions. This view fell into line with a general emphasis on the systems concept in the study of Earth. As geomorphology evolved, it became more interdisciplinary, as we shall see. This, too, was part of an overall trend in the earth sciences toward an approach that viewed subjects in broad, cross-disciplinary terms as opposed to a narrow focus on specific areas of study.

Landforms and Processes

Two concerns are foremost within the realm of geomorphology, and these concerns reflect the stages of its history. First, in line with Davis's original conception of geomorphology as an area of science devoted to classifying and describing natural features, there is its concern with topography. The latter may be defined as the configuration of Earth's surface, including its relief (elevation and other in equalities) as well as the position of physical features.

These physical features are called landforms, examples of which include mountains, plateaus, and valleys. Geomorphology always has involved classification, and early scientists working in this subdiscipline addressed the classification of landforms. Other systems of classification, however, are not so concerned with cataloging topographical features themselves as with differentiating the processes that shaped them. This brings us to the other area of interest in geomorphology: the study of how landforms came into being.

Shaping the Earth

Among the processes that drive the shaping of landforms is plate tectonics, or the shifting of large, movable segments of lithosphere (the crust and upper layer of Earth's mantle).

Other processes also shape landforms. Included among these processes are weathering, the breakdown of rocks and minerals at or near the surface of Earth due to physical or chemical processes; erosion, the movement of soil and rock due to forces produced by water, wind, glaciers, gravity, and other influences; and mass wasting or mass movement, the transfer of earth material, by processes that include flow, slide, fall, and creep, down slopes. Also of interest are fluvial and eolian processes (those that result from water flow and wind, respectively) as well as others related to glaciers and coastal formations.

Human activity also can play a significant role in shaping Earth. This effect may be direct, as when the construction of cities, the building of dams, or the excavation of mines alters the landscape. On the other hand, it can be indirect. In the latter instance, human activity in the biosphere exerts an impact, as when the clearing of forest land or the misuse of crop land results in the formation of a dust bowl.

Interdisciplinary Studies

Geomorphology is characteristic of the earth sciences as a whole in its emphasis on an interdisciplinary approach. As is true of earth scientists in general, those studying landforms and the processes that shape them do not work simply in one specialty. Among the areas of interest in geomorphology are, for example, deep-sea geomorphology, which draws on oceanography, and planetary geomorphology, the study of landscapes on other planets.

When studying coastal geomorphology, a geologist may draw on realms as diverse as fluid mechanics (an area of physics that studies the behavior of gases and liquids at rest and in motion) and sedimentology. The investigation of such processes as erosion and mass wasting calls on knowledge in the atmospheric sciences as well as the physics and chemistry of soil. It is almost inevitable that a geomorphologic researcher will draw on geophysics as well as on such subspecialties as volcanology. These studies may go beyond the "hard sciences," bringing in such social sciences as geography.

Real-life Applications

Subsidence

Subsidence refers to the process of subsiding (settling or descending), on the part of either an air column or the solid earth, or, in the case of solid earth, to the resulting formation or depression. Subsidence in the atmosphere is discussed briefly in the entry Convection. Subsidence that occurs in the solid earth, known as geologic subsidence, is the settling or sinking by a body of rock or sediment. (The latter can be defined as material deposited at or near Earth's surface from a number of sources, most notably preexisting rock).

Many geomorphologic processes can be caused either by nature or by human beings. An example of natural subsidence takes place in the aftermath of an earthquake, during which large areas of solid earth may simply drop by several feet. Another example can be observed at the top of a volcano some time after it has erupted, when it has expelled much of its material (i.e., magma) and, as a result, has collapsed.

Natural subsidence also may result from cave formation in places where underground water has worn away limestone. If the water erodes too much limestone, the ceiling of the cave will subside, usually forming a sinkhole at the surface. The sinkhole may fill with water, making a lake; the formation of such sinkholes in many spots throughout an area (whether the sinkholes become lakes or not), is known as karst topography.

In places where the bedrock is limestone—particularly in the sedimentary basins of rivers—karst topography is likely to develop. The United States contains the most extensive karst region in the world, including the Mammoth cave system in Kentucky. Karst topography is very pronounced in the hills of southern China, and karst landscapes have been a prominent feature of Chinese art for centuries. Other extensive karst regions can be found in southern France, Central America, Turkey, Ireland, and England.

Man-made Subsidence

Man-made subsidence often ensues from the removal of groundwater or fossil fuels, such as petroleum or coal. Groundwater removal can be perfectly safe, assuming the area experiences sufficient rainfall to replace, or recharge, the lost water. If recharging does not occur in the necessary proportions, however, the result will be the eventual collapse of the aquifer, a layer of rock that holds groundwater.

In so-called room-and-pillar coal mining, pillars, or vertical columns, of coal are left standing, while the areas around them are extracted. This method maintains the ceiling of the "room" that has been mined of its coal. After the mine is abandoned, however, the pillar eventually may experience so much stress that it breaks, leading to the collapse of the mined room. As when the ceiling of a cave collapses, the subsidence of a coal mine leaves a visible depression above ground.

Uplift

As its name implies, uplift describes a process and results opposite to those of subsidence. In uplift the surface of Earth rises, owing either to a decrease in downward force or to an increase in upward force. One of the most prominent examples of uplift is seen when plates collide, as when India careened into the southern edge of the Eurasian landmass some 55 million years ago. The result has been a string of mountain ranges, including the Himalayas, Karakoram Range, and Hindu Kush, that contain most of the world's tallest peaks.

Plates move at exceedingly slow speeds, but their mass is enormous. This means that their inertia (the tendency of a moving object to keep moving unless acted upon by an outside force) is likewise gargantuan in scale. Therefore, when plates collide, though they are moving at a rate equal to only a few inches a year, they will keep pushing into each other like two automobiles crumpling in a head-on collision. Whereas a car crash is over in a matter of seconds, however, the crumpling of continental masses takes place over hundreds of thousands of years.

When sea floor collides with sea floor, one of the plates likely will be pushed under by the other one, and, likewise, when sea floor collides with continental crust, the latter will push the sea floor under. This results in the formation of volcanic mountains, such as the Andes of South America or the Cascades of the Pacific Northwest, or volcanic islands, such as those of Japan, Indonesia, or Alaska's Aleutian chain.

Isostatic Compensation

In many other instances, collision, compression, and extension cause uplift. On the other hand, as noted, uplift may result from the removal of a weight. This occurs at the end of an ice age, when glaciers as thick as 1.9 mi. (3 km) melt, gradually removing a vast weight pressing down on the surface below.

This movement leads to what is called isostatic compensation, or isostatic rebound, as the crust pushes upward like a seat cushion rising after a person is longer sitting on it. Scandinavia is still experiencing uplift at a rate of about 0.5 in. (1 cm) per year as the after-effect of glacial melting from the last ice age. The latter ended some 10,000 years ago, but in geologic terms this is equivalent to a few minutes' time on the human scale.

Islands

Geomorphology, as noted earlier, is concerned with landforms, such as mountains and volcanoes as well as larger ones, including islands and even continents. Islands present a particularly interesting area of geomorphologic study. In general, islands have certain specific characteristics in terms of their land structure and can be analyzed from the standpoint of the geosphere, but particular islands also have unique ecosystems, requiring an interdisciplinary study that draws on botany, zoology, and other subjects.

In addition, there is something about an island that has always appealed to the human imagination, as evidenced by the many myths, legends, and stories about islands. Some examples include Homer's Odyssey, in which the hero Odysseus visits various islands in his long wanderings; Thomas More's Utopia, describing an idealized island republic; Robinson Crusoe, by Daniel Defoe, in which the eponymous hero lives for many years on an island with no companion but the trusty native Friday; Treasure Island, by Robert Louis Stevenson, in which the island is the focus of a treasure hunt; and Mark Twain's Adventures of Huckleberry Finn, depicting Jackson Island in the Mississippi River, to which Huckleberry Finn flees to escape "civilization."

One of the favorite subjects of cartoonists is that of a castaway stranded on a desert island, a mound of sand with no more than a single tree. Movies, too, have long portrayed scenarios, from the idyllic to the brutal, that take place on islands, particularly deserted ones, a notable example being Cast Away. A famous line by the English poet John Donne warns that "no man is an island," implying that many wish they could enjoy the independence suggested by the concept of an island. Within the Earth system, however, nothing is fully independent, and, as we shall see, this is certainly the case where islands are concerned.

The Islands of Earth

Earth has literally tens of thousands of islands. Just two archipelagos (island chains), those that make up the Philippines and Indonesia, include thousands of islands each. While there are just a few dozen notable islands on Earth, many more dot the planet's seas and oceans. The largest are these:

- Greenland (Danish, northern Atlantic): 839,999 sq. mi.(2,175,597 sq km).

- New Guinea (divided between Indonesia and Papua New Guinea, western Pacific): 316,615 sq. mi. (820,033 sq km).
- Borneo (divided between Indonesia and Malaysia, western Pacific): 286,914 sq. mi. (743,107 sq km).
- Madagascar (Malagasy Republic, western Indian Ocean): 226,657 sq. mi. (587,042 sq km).
- Baffin (Canadian, northern Atlantic): 183,810 sq. mi. (476,068 sq km).
- Sumatra (Indonesian, northeastern Indian Ocean): 182,859 sq. mi. (473,605 sq km).

The list could go on and on, but it stops at Sumatra because the next-largest island, Honshu (part of Japan), is less than half as large, at 88,925 sq. mi. (230,316 sq km). Clearly, not all islands are created equal, and though some are heavily populated or enjoy the status of independent nations (e.g., Great Britain at number eight or Cuba at number 15), they are not necessarily the largest. On the other hand, some of the largest are among the most sparsely populated.

Of the 32 largest islands in the world, more than a third are in the icy northern Atlantic and Arctic, with populations that are small or practically nonexistent. Greenland's population, for instance, was just over 59,000 in 1998, while that of Baffin Island was about 13,200. On both islands, then, each person has about 14 frozen sq. mi. (22 sq km) to himself or herself, making them among the most sparsely populated places on Earth.

Continents, Oceans and Islands

Australia, of course, is not an island but a continent, a difference that is not related directly to size. If Australia were an island, it would be by far the largest. Australia is regarded as a continent, however, because it is one of the principal landmasses of the Indo-Australian plate, which is among a handful of major continental plates on Earth. Whereas continents are more or less permanent (though they have experienced considerable rearrangement over the eons), islands come and go, seldom lasting more than 10 million years. Erosion or rising sea levels remove islands, while volcanic explosions can create new ones, as when an eruption off the coast of Iceland resulted in the formation of an island, Surtsey, in 1963.

Islands are of two types, continental and oceanic. Continental islands are part of continental shelves (the submerged, sloping ledges of continents) and may be formed in one of two ways. Rising ocean waters either cover a coastal region, leaving only the tallest mountains exposed as islands or cut off part of a peninsula, which then becomes an island. Most of Earth's significant islands are continental and are easily spotted as such, because they lie at close proximity to continental landmasses. Many other continental islands are very small, however; examples include the barrier islands that line the East Coast of the United States. Formed from mainland sand brought to the coast by rivers, these are technically not continental islands, but they more clearly fit into that category than into the grouping of oceanic islands.

Oceanic islands, of which the Hawaiian-Emperor island chain and the Aleutians off the Alaskan coast are examples, form as a result of volcanic activity on the ocean floor. In most cases, there is a region of high volcanic activity, called a hot spot, beneath the plates, which move across the hot spot. This is the situation in Hawaii, and it explains why the volcanoes on the southern islands are still active while those to the north are not: the islands themselves are moving north across the hot spot. If two plates converge and one subducts, a deep trench with a parallel chain of volcanic islands may develop. Exemplified by the Aleutians, these chains are called island arcs.

Island Ecosystems

The ecosystem, or community of all living organisms, on islands can be unique owing to their separation from continents. The number of life-forms on an island is relatively small and can encompass some unusual circumstances compared with the larger ecosystems of continents. Ireland, for instance, has no native snakes, a fact "explained" by the legend that Saint Patrick drove them away. Hawaii and Iceland are also blessedly free of serpents.

Oceanic islands, of course, tend to have more unique ecosystems than do continental islands. The number of land-based animal life-forms is necessarily small, whereas the varieties of birds, flying insects, and surrounding marine life will be greater owing to those creatures' mobility across water. Vegetation is relatively varied, given the fact that winds, water currents, and birds may carry seeds.

Nonetheless, ecosystems of islands tend to be fairly delicate and can be upset by the human introduction of new predators (e.g., dogs) or new creatures to consume plant life (e.g., sheep). These changes sometimes can have disastrous effects on the overall balance of life on islands. Overgrazing may even open up the possibility of erosion, which has the potential of bringing an end to an island's life.

Biogeomorphology

Slope stabilization by Chilean rhubarb on the coasts of Chacao Channel.
Vegetation have mostly a protective effect on slopes.

Biogeomorphology and ecogeomorphology are the study of interactions between organisms and the development of landforms, and are thus fields of study within geomorphology and ichnology. Organisms affect geomorphic processes in a variety of ways. For example, trees can reduce landslide potential where their rootspenetrate to underlying rock, plants and their litter inhibit soil erosion, biochemicals produced by plants accelerate the chemical weathering of bedrock and regolith, and marine animals cause the bioerosion of coral. The study of the interactions between marine biota and coastal landform processes is called coastal biogeomorphology.

Phytogeomorphology is an aspect of biogeomorphology that deals with the narrower subject of how terrain affects plant growth. In recent years a large number of articles have appeared in the literature dealing with how terrain attributes affect crop growth and yield in farm fields, and while they don't use the term phytogeomorphology the dependencies are the same. Precision agriculture models where crop variability is at least partially defined by terrain attributes can be considered as phytogeomorphological precision agriculture.

Biogeomorphology is a multidisciplinary focus of geomorphology that takes research approaches from both geomorphology and ecology. It is a sub discipline of geomorphology. Biogeomorphology can be synthesized into two distinct approaches:

1. The influences that geomorphology plays on the biodiversity and distribution of flora and fauna.
2. The influences that biotic factors have on the way landforms are developed.

There has been much work on these approaches such as; the effect that parent material has on the distribution of plants,the increase of precipitation due to an influx of transpiration, the stability of a hillslope due to the abundance of vegetation or, the increase of sedimentation due to a beaver dam. Biogeomorphology shows the axiomatic relationship between certain land forming processes and biotic factors. That is, certain geomorphic processes shape the biota and biotic factors can shape land forming processes.

Darwin's work on earthworms, the caption of which reads "A tower-like casting, probably ejected by a species of Perichæta, from the Botanic Garden."

The earliest work related to biogeomorphology was Charles Darwin's 1881 book titled *The Formation of Vegetable Mould through the Action of Worms*. Although the field of biogeomorphology had not yet been named, Darwin's work represents the earliest examination of a faunal organism influencing landscape process and form. Charles Darwin begins his work on worms with an examination of behavior and physiology, which then moves towards topics related to geomorphology, pedogenesis, and bioturbation.Observations and measurements of soil moved by earthworms, and emphasis on the role of earthworms in formation of humus, fertility of soils, and mixing of soils were all described in the book, which began to change the perspective on earthworms from pest to critical agent of pedogenesis. Despite the popularity of Darwin's final work, the scientific community was slow to recognize the significance of examining the role of organisms in influencing landscapes.

It wasn't until the late twentieth century that biogeomorphology began attracting the attention of more than a handful of researchers.

Research Approaches

There are two approaches to research in biogeomorphology. One is through the statistical and empirically derived means. This is an approach commonly used in the fields of ecologyand biology. The approach is simply to employ large replication studies and deriving patterns from statistical data.Whereas taking a more geomorphic research approach tends to derive patterns via theoretic knowledge and detailed measurements of multiple factors. In turn, this uses smaller sample sizes than that of large replication studies.

Biogeomorphological Processes

There are several biogeomorphological processes. Bioerosion is the weathering and removal of abiotic material via organic processes. This can either be passive or active. Moreover, bioerosion is the chemical and or the mechanical weathering of landforms due to organic means. Bioprotection is essentially the effect that organisms have on reducing the action of geomorphic processes. Best shown by algae covering a rock surface acts as a buffer from the waves erosive work. Bioconstruction refers to bioconstructors or ecosystem engineers. Ecosystem engineers are organisms that build mounds, dams, reefs &c. More specifically they are organisms that change the environment physically, directly or indirectly controlling the abundance of resources available in the environment accessible to organisms.

Themes of Complex Systems in Biogeomorphology

There are four main themes that underline the complex systems within biogeomorphology. The first of which is multiple causality. Multiple casualty is the way in which biota is deposited. More specifically, multiple casualty is caused by various processes. That is, processes such as fires, floods and hillslope instability directly or indirectly determining the distribution of flora and in turn fauna. Ecosystem engineers are another theme underlying complex system of biogeomorphology. These organisms have the most profound effect on the overall ecosystem structure. Some of the most common ecosystem engineers are earthworms. Earthworms aid in the production of humus and increase both soil aeration and area for roots and root hairs to utilize. With more space for roots, this can increase soil stability. Another strong example of ecosystem engineers are beavers. Beavers can increase the sedimentation in a channel as well as increase runoff rates due to a reduction of vegetative cover needed to build their dams. Ecological topology is another theme of complex systems in biogeomorphology. This theme focuses on how the biota varies based on geographic location. This ecological topology is controlled by a concept called stability domain. Stability domain describes the interaction of a set species and certain abiotic factors that act as a medium to the function and structure of an environment. The final of the four underlying themes of the complex systems in biogeomorphology is ecological memory. Ecological memory is where certain biotic and abiotic factors have a recursive relationship and therefore can be encoded in organisms and the immediate environment. An example of this can simply be the flame retardant properties in the bark of Coastal Sequoias due to the recurrence of fires.

Climate Change

Biogeomorphology and ecogeomorphology can aid with assessing the impacts of global climate change. This can especially be seen in coastal and estuarine systems due to; sea level rise, increased global temperatures, increased sea temperature, a higher frequency in and intensity of storms, and varying distributions of precipitation. Biogeomorphology can outline some of the effects of climate change due to biocomplexity. Biocomplexity is in reference to the complex way in which organisms interact with their environment and its effect on biodiversity. Using statistical data, one can derive how these environmental changes will affect the biodiversity of different trophic levels and different keystone species.

Aeolian Processes

Wind erosion of soil at the foot of Chimborazo, Ecuador.

Rock carved by drifting sand below Fortification Rock in Arizona.

Aeolian processes, pertain to wind activity in the study of geology and weather and specifically to the wind's ability to shape the surface of the Earth (or other planets). Winds may erode, transport, and deposit materials and are effective agents in regions with sparse vegetation, a lack of soil moisture and a large supply of unconsolidated sediments. Although water is a much more powerful eroding force than wind, aeolian processes are important in arid environments such as deserts.

The term is derived from the name of the Greek god Aeolus, the keeper of the winds.

Wind Erosion

A rock sculpted by wind erosion in the Altiplano region of Bolivia.

Wind erodes the Earth's surface by deflation (the removal of loose, fine-grained particles by the turbulent action of the wind) and by abrasion (the wearing down of surfaces by the grinding action and sandblasting by windborne particles).

Sand blowing off a crest in the Kelso Dunes of the Mojave Desert, California.

Regions which experience intense and sustained erosion are called deflation zones. Most aeolian deflation zones are composed of desert pavement, a sheet-like surface of rock fragments that remains after wind and water have removed the fine particles. Almost half of Earth's desert surfaces are stony deflation zones. The rock mantle in desert pavements protects the underlying material from deflation.

A dark, shiny stain, called desert varnish or rock varnish, is often found on the surfaces of some desert rocks that have been exposed at the surface for a long period of time. Manganese, iron oxides, hydroxides, and clay minerals form most varnishes and provide the shine.

Wind-carved alcove in the Navajo Sandstone near Moab, Utah.

Deflation basins, called blowouts, are hollows formed by the removal of particles by wind. Blowouts are generally small, but may be up to several kilometers in diameter.

Wind-driven grains abrade landforms. In parts of Antarctica wind-blown snowflakes that are technically sediments have also caused abrasion of exposed rocks.Grinding by particles carried in the wind creates grooves or small depressions. Ventifacts are rocks which have been cut, and sometimes polished, by the abrasive action of wind.

Sculpted landforms, called yardangs, are up to tens of meters high and kilometers long and are forms that have been streamlined by desert winds. The famous Great Sphinx of Giza in Egypt may be a modified yardang.

Transport

Particles are transported by winds through suspension, saltation (skipping or bouncing) and creeping (rolling or sliding) along the ground.

Dust storm approaching Spearman, Texas April 14, 1935.

Small particles may be held in the atmosphere in suspension. Upward currents of air support the weight of suspended particles and hold them indefinitely in the surrounding air. Typical winds near Earth's surface suspend particles less than 0.2 millimeters in diameter and scatter them aloft as dust or haze.

Saltation is downwind movement of particles in a series of jumps or skips. Saltation normally lifts sand-size particles no more than one centimeter above the ground and proceeds at one-half to one-third the speed of the wind. A saltating grain may hit other grains that jump up to continue the saltation. The grain may also hit larger grains that are too heavy to hop, but that slowly creep forward as they are pushed by saltating grains. Surface creep accounts for as much as 25 percent of grain movement in a desert.

Dust storm in Amarillo, Texas.

Aeolian turbidity currents are better known as dust storms. Air over deserts is cooled significantly when rain passes through it. This cooler and denser air sinks toward the desert surface. When it reaches the ground, the air is deflected forward and sweeps up surface debris in its turbulence as a dust storm.

Crops, people, villages, and possibly even climates are affected by dust storms. Some dust storms are intercontinental, a few may circle the globe, and occasionally they may engulf entire planets. When the Mariner 9 spacecraft entered its orbit around Mars in 1971, a dust storm lasting one month covered the entire planet, thus delaying the task of photo-mapping the planet's surface.

Most of the dust carried by dust storms is in the form of silt-size particles. Deposits of this windblown silt are known as loess. The thickest known deposit of loess, 335 meters, is on the Loess Plateau in China. This very same Asian dust is blown for thousands of miles, forming deep beds in places as far away as Hawaii. In Europeand in the Americas, accumulations of loess are generally from 20 to 30 meters thick. The soils developed on loess are generally highly productive for agriculture.

A massive sand storm cloud is about to envelop a military camp as it rolls over Al Asad, Iraq, just before nightfall on April 27, 2005.

Aeolian transport from deserts plays an important role in ecosystems globally, e.g. by transport of minerals from the Sahara to the Amazon basin. Saharan dust is also responsible for forming red clay soils in southern Europe. Aeolian processes are affected by human activity, such as the use of 4x4 vehicles.

Small whirlwinds, called dust devils, are common in arid lands and are thought to be related to very intense local heating of the air that results in instabilities of the air mass. Dust devils may be as much as one kilometer high.

Wind-deposited materials hold clues to past as well as to present wind directions and intensities. These features help us understand the present climate and the forces that molded it. Wind-deposited sand bodies occur as sand sheets, ripples, and dunes.

Sand sheets are flat, gently undulating sandy plots of sand surfaced by grains that may be too large for saltation. They form approximately 40 percent of aeolian depositional surfaces. The Selima Sand Sheet in the eastern Sahara Desert, which occupies 60,000 square kilometers in southern Egypt and northern Sudan, is one of the Earth's largest sand sheets. The Selima is absolutely flat in a few places; in others, active dunes move over its surface.

Wind blowing on a sand surface ripples the surface into crests and troughs whose long axes are perpendicular to the wind direction. The average length of jumps during saltation corresponds to the wavelength, or distance between adjacent crests, of the ripples. In ripples, the coarsest materials collect at the crests causing inverse grading. This distinguishes small ripples from dunes, where the coarsest materials are generally in the troughs. This is also a distinguishing feature between water laid ripples and aeolian ripples.

Accumulations of sediment blown by the wind into a mound or ridge, dunes have gentle upwind slopes on the windward side. The downwind portion of the dune, the lee slope, is commonly a steep avalanche slope referred to as a slipface. Dunes may have more than one slipface. The minimum height of a slipface is about 30 centimeters.

Wind-blown sand moves up the gentle upwind side of the dune by saltation or creep. Sand accumulates at the brink, the top of the slipface. When the buildup of sand at the brink exceeds the angle of repose, a small avalanche of grains slides down the slipface. Grain by grain, the dune moves downwind.

Some of the most significant experimental measurements on aeolian sand movement were performed by Ralph Alger Bagnold, a British army engineer who worked in Egypt prior to World War II. Bagnold investigated the physics of particles moving through the atmosphere and deposited by wind. He recognized two basic dune types, the crescentic dune, which he called "barchan", and the linear dune, which he called longitudinal or "seif" (Arabic for "sword").

A 2011 study published in *Catena* examined the effect of vegetation on aeolian dust accumulation in the semiarid steppe of northern China. Using a series of trays with different vegetation coverage and a control model with none, the authors found that an increase in vegetation coverage improves the efficiency of dust accumulation and adds more nutrients to the environment, particularly organic carbon. Two critical point were revealed by their data: 1. the efficiency of trapping dust increases slowly above 15% coverage, and decreases rapidly below 15% coverage. 2. at around 55%-75% coverage, dust accumulation reaches a maximum capacity.

In Europe, the European Commission requested the Joint Research Centre to develop the first pan-European wind erosion map. In a first step, a group of scientists have used the LUCAS topsoil dataset to develop the wind erosion susceptibility of European soils. Then, they have developed an index to land susceptibility for making a qualitative assessment of wind erosion. Finally, they modified the RWEQ model to estimate the soil Loss Due to Wind Erosion in European Agricultural Soils.

A three-year quantitative study on the effects of vegetation removal on wind erosion found that the removal of grasses in an aeolian environment increased the rate of soil deposition. In the same study, a relationship was shown between decreasing plant density with decreasing soil nutrients. Similarly, horizontal soil flux across the test site was shown to increase with increasing vegetation removal.

A 1998 study published in *Earth Surfaces Processes and Landforms* investigated the relationship between vegetative cover on sand surfaces with the rate of sand transport. It was found that sand flux decreased exponentially with vegetation cover. This was done by measuring plots of land with varying degrees of vegetation against rates of sand transport. The authors contend that this relationship can be utilized to manipulate rates of sediment flux by introducing vegetation in an area or to quantify human impact by recognizing vegetation loss's effect on sandy landscapes.

Mesquite Flat Dunes in Death Valleylooking toward the Cottonwood Mountains from the north west arm of Star Dune.

Holocene eolianite deposit on Long Island, The Bahamas. This unit is formed of wind-blown carbonate grains.

Deep, eroding glaciofluvial deposits alongside the Matanuska River, Alaska.

In geography and geology, fluvial processes are associated with rivers and streams and the deposits and landforms created by them. When the stream or rivers are associated with glaciers, ice sheets, or ice caps, the term glaciofluvial or fluvioglacial is used.

Fluvial Processes

Fluvial processes include the motion of sediment and erosion or deposition on the river bed.

Erosion by moving water can happen in two ways. Firstly, the movement of water across the stream bed exerts a shear stress directly onto the bed. If the cohesive strength of the substrate is lower than the shear exerted, or the bed is composed of loose sediment which can be mobilized by such stresses, then the bed will be lowered purely by clearwater flow. However, if the river carries significant quantities of sediment, this material can act as tools to enhance wear of the bed (abrasion). At the same time the fragments themselves are ground down, becoming smaller and more rounded (attrition).

Sediment in rivers is transported as either bedload (the coarser fragments which move close to the bed) or suspended load (finer fragments carried in the water). There is also a component carried as dissolved material.

For each grain size there is a specific velocity at which the grains start to move, called *entrainment velocity*. However the grains will continue to be transported even if the velocity falls below the entrainment velocity due to the reduced (or removed) friction between the grains and the river bed. Eventually the velocity will fall low enough for the grains to be deposited. This is shown by the Hjulström curve.

A river is continually picking up and dropping solid particles of rock and soil from its bed throughout its length. Where the river flow is fast, more particles are picked up than dropped. Where the river flow is slow, more particles are dropped than picked up. Areas where more particles are dropped are called alluvial or flood plains, and the dropped particles are called alluvium.

Even small streams make alluvial deposits, but it is in the flood plains and deltas of large rivers that large, geologically-significant alluvial deposits are found.

The amount of matter carried by a large river is enormous. The names of many rivers derive from the color that the transported matter gives the water. For example, the Huang He in China is literally translated "Yellow River", and the Mississippi River in the United States is also called "the Big Muddy". It has been estimated that the Mississippi River annually carries 406 million tons of sediment to the sea, the Yellow River 796 million tons, and the Po River in Italy 67 million tons.

Erosion and Tectonics

The interaction between erosion and tectonics has been a topic of debate since the early 1990s. While the tectonic effectson surface processes such as erosion have long been recognized (for example, river formation as a result of tectonic uplift), the opposite (erosional effects on tectonic activity) has only recently been addressed. The primary questions surrounding this topic are what types of interactions exist between erosion and tectonics and what are the implications of these interactions. While this is still a matter of debate, one thing is clear, the Earth's landscape is a product of two factors: tectonics, which can create topography and maintain relief through surface and rock uplift, and climate, which mediates the erosional processes that wear away upland areas over time. The interaction of these processes can form, modify, or destroy geomorphic features on the Earth's surface.

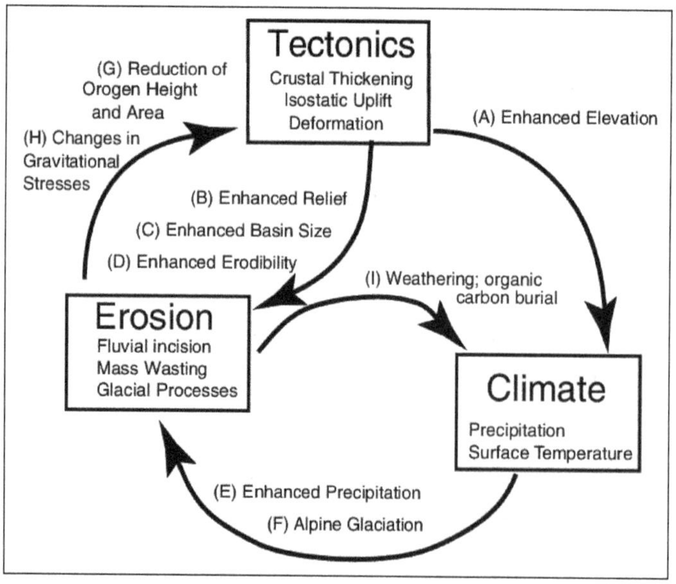

Interactions and feedback pathways for tectonics and erosional processes.

Tectonic Processes

The term tectonics refers to the study of the Earth's surface structure and the ways in which it changes over time. Tectonic processes typically occur at plate boundaries which are one of three types: convergent boundaries, divergent boundaries, or transform boundaries. These processes form and modify the topography of the Earth's surface, effectively increasing relief through the mechanisms of isostatic uplift, crustal thickening, and deformation in the form of faulting and folding. Increased elevations, in relation to regional base levels, lead to steeper river channel gradients and an increase in orographically localized precipitation, ultimately resulting in drastically increased erosion rates. The topography, and general relief, of a given area determines the velocity at which surface runoff will flow, ultimately determining the potential erosive power of the runoff. Longer, steeper slopes are more prone to higher rates of erosion during periods of heavy rainfall than shorter, gradually sloping areas. Thus, large mountain ranges, and other areas of high relief, formed through tectonic uplift will have significantly higher rates of erosion. Additionally, tectonics can directly influence erosion rates on a short timescale, as is clear in the case of earthquakes, which can trigger landslides and weaken surrounding rock through seismic disturbances.

While tectonic uplift in any case will lead to some form of increased elevation, thus higher rates of erosion, a primary focus is set on isostatic uplift as it provides a fundamental connection between the causes and effects of erosional-tectonic interactions.

Isostatic Uplift

Understanding the principle of isostasy is a key element to understanding the interactions and feedbacks shared between erosion and tectonics. The principle of isostasy states that when free to move vertically, lithosphere floats at an appropriate level in the asthenosphere so that the pressure at a depth of compensation in the asthenosphere well below the base of the lithosphere is the same. Isostatic uplift is both a cause and an effect of erosion. When deformation occurs in the form of crustal thickening an isostatic response is induced causing the thickened crust to sink, and surrounding thinner crust to uplift. The resulting surface uplift leads to enhanced elevations, which in turn induces erosion. Alternatively, when a large amount of material is eroded away from the Earth's surface uplift occurs in order to maintain isostatic equilibrium. Because of isostasy, high erosion rates over significant horizontal areas can effectively suck up material from the lower crust and/or upper mantle. This process is known as *isostatic rebound* and is analogous to Earth's response following the removal of large glacial ice sheets.

Isostatic uplift and corresponding erosion are responsible for the formation of regional-scale geologic features as well as localized structures. Two such examples include:

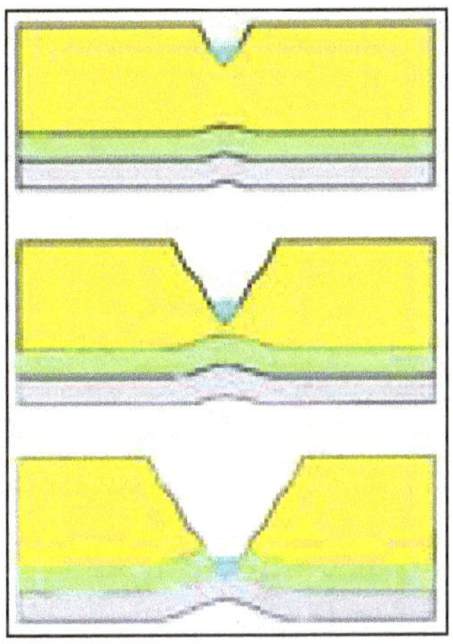

Formation of a river anticline.

- Continental shields- Generally large areas of low relief (<100 m) in the Earth's crust where

Precambrian crystalline igneous and high-grade metamorphic rocks are exposed. Shields are considered tectonically stable areas in comparison to the activity occurring at their margins and the boundaries between plates, but their formation required large amounts of tectonic activity and erosion. Shields, along with stable platforms, are the basic tectonic components of continents, therefore understanding their development is critical to understanding the development of other surface features on Earth. Initially, a mountain belt is formed at a convergent plate margin. Transformation of a mountain belt to a shield is majorly dependent on two factors: (1) erosion of the mountain belt by running water and (2) isostatic adjustment resulting from the removal of surface rock due to erosion. This process of erosion followed by isostatic adjustment continues until the system is at isostatic equilibrium. At this point large-scale erosion can no longer occur because the surface has eroded down to nearly sea-level and uplift ceases due to the system's state of equilibrium.

- River anticlines- Geologic structures formed through the focused uplift of rock underlying confined areas of high erosion (i.e., rivers). Isostatic rebound resulting from the rapid removal of overlying rock, via erosion, causes the weakened areas of crustal rock to uplift from the apex of the river. In order for the development of these structures to occur the erosion rate of the river must exceed both the average erosional rate of the area, and the rate of uplift of the orogen. The two factors influencing the development of these structures are stream power of the associated river and the flexural rigidity of the crust in the area. The combination of increased stream power with decreased flexural rigidity results in the system's progression from a transverse anticline to a river anticline.

Channel Flow

Channel flow describes the process through which hot, viscous crustal material flows horizontally between the upper crust and lithospheric mantle, and is eventually pushed to the surface. This model aims to explain features common to metamorphic hinterlands of some collisional orogens, most notably the Himalaya–Tibetan plateau system. In mountainous areas with heavy rainfall (thus, high erosion rates) deeply incising rivers will form. As these rivers wear away the Earth's surface two things occur: (1) pressure is reduced on the underlying rocks effectively making them weaker and (2) the underlying material moves closer to the surface. This reduction of crustal strength, coupled with the erosional exhumation, allows for the diversion of the underlying channel flow toward the Earth's surface.

Erosional Processes

The term erosion refers to the group of natural processes, including weathering, dissolution, abrasion, corrosion, and transportation, by which material is worn away from the Earth's surface to be transported and deposited in other locations.

- Differential erosion- Erosion that occurs at irregular or varying rates, caused by the differences in the resistance and hardness of surface materials; softer and weaker rocks are rapidly worn away, whereas harder and more resistant rocks remain to form ridges, hills, or mountains. Differential erosion, along with the tectonic setting, are two of the most important controls on the evolution of continental landscapes on Earth.

A natural arch produced by erosion of differentially weathered rock in Jebel Kharaz (Jordan).

The feedback of erosion on tectonics is given by the transportation of surface, or near-surface, mass (rock, soil, sand, regolith, etc.) to a new location. This redistribution of material can have profound effects on the state of gravitational stresses in the area, dependent on the magnitude of mass transported. Because tectonic processes are highly dependent on the current state of gravitational stresses, redistribution of surface material can lead to tectonic activity. While erosion in all of its forms, by definition, wears away material from the Earth's surface, the process of mass wasting as a product of deep fluvial incision has the highest tectonic implications.

Mass Wasting

Talus cones produced by mass wasting, north shore of Isfjord, Svalbard, Norway.

Mass wasting is the geomorphic process by which surface material move downslope typically as a mass, largely under the force of gravity As rivers flow down steeply sloping mountains, deep channel incision occurs as the river's flow wears away the underlying rock. Large channel incision progressively decreases the amount of gravitational force needed for a slope failure event to occur, eventually resulting in mass wasting. Removal of large amounts of surface mass in this fashion will induce an isostatic response resulting in uplift until equilibrium is reached.

Effects on Structural Evolution

Recent studies have shown that erosional and tectonic processes have an effect on the structural evolution of some geologic features, most notably orogenic wedges. Highly useful sand box models, in which horizontal layers of sand are slowly pressed against a backstop, have shown that the geometries, structures, and kinematics of orogenic wedge formation with and without erosion and sedimentation are significantly different. Numerical models also show that the evolution of orogens, their final tectonic structure, and the potential development of a high plateau, all are sensitive to the long term climate over the mountains, for example, the concentration of precipitation in one side of the orogen due to orographic lift under a dominant wind direction.

Tectonic Uplift

Kupe's Sail at Palliser Bay in New Zealand.

Tectonic uplift is the portion of the total geologic uplift of the mean Earth surface that is not attributable to an isostatic response to unloading. While isostatic response is important, an increase in the mean elevation of a region can only occur in response to tectonic processes of crustal thickening (such as mountain building events), changes in the density distribution of the crust and underlying mantle, and flexural support due to the bending of rigid lithosphere.

One should also take into consideration the effects of denudation (processes that wear away the earth's surface). Within the scope of this topic, uplift relates to denudation in that denudation brings buried rocks closer to the surface. This process can redistribute large loads from an elevated region to a topographically lower area as well – thus promoting an isostatic response in the region of denudation (which can cause local bedrock uplift). The timing, magnitude, and rate of denudation can be estimated by geologists using pressure-temperature studies.

Crustal Thickening

Crustal thickening has an upward component of motion and often occurs when continental crust is thrust onto continental crust. Basically nappes (thrust sheets) from each plate collide and begin to stack one on top of the other; evidence of this process can be seen in preserved ophiolitic nappes (preserved in the Himalaya), and in rocks with an inverted metamorphic gradient. The preserved inverted metamorphic gradient indicates that nappes were actually stacked on top of each other so quickly, that hot rocks did not have time to equilibrate before being thrust on top of cool rocks. The process of nappe stacking can only continue for so long, as gravity will eventually disallow further vertical growth (there is an upper limit to vertical mountain growth).

Density Distribution of the Crust and Underlying Mantle

Although the raised surfaces of mountain ranges mainly result from crustal thickening, there are other forces at play that are responsible for the tectonic activity. All tectonic processes are driven by gravitational force when density differences are present. A good example of this would be the large-scale circulation of the Earth's mantle. Lateral density variations near the surface (such as the creation, cooling, and subduction of oceanic plates) also drive plate motion.

The dynamics of mountain ranges are governed by differences in the gravitational potential energy of entire *columns* of the lithosphere. If a change in surface height represents an isostatically compensated change in crustal thickness, the rate of change of potential energy per unit surface area is proportional to the rate of increase of average surface height. The highest rates of working against gravity are required when the thickness of the crust (not the lithosphere) changes.

Lithospheric Flexure

Lithosphere on the oceanward side of an oceanic trench at a subduction zone will curve upwards due to the elastic properties of the Earth's crust.

Orogenic Uplift

Orogenic uplift is the result of tectonic-plate collisions and results in mountain ranges or a more modest uplift over a large region. Perhaps the most extreme form of orogenic uplift is a continental-continental crustal collision. In this process, two continents are sutured together and large mountain ranges are produced. The collision of the Indian and Eurasian plates is a good example of the extent to which orogenic uplift can reach. Heavy thrust faulting (of the Indian plate beneath the Eurasian plate) and folding are responsible for the suturing together of the two plates. The collision of the Indian and Eurasian plates not only produced the Himalaya but is also responsible for crustal thickening north into Siberia. The Pamir Mountains, Tian Shan, Altai, Hindu Kush, and other mountain belts are all examples of mountain ranges formed in response to the collision of the Indian with the Eurasian plate. Deformation of continental lithosphere can take place in several possible modes.

The Ozark Plateau is a broad uplifted area which resulted from the Permian Ouachita Orogeny to the south in the states of Arkansas, Oklahoma and Texas. Another related uplift is the Llano Uplift in Texas, a geographical location named after its uplift features.

The Colorado Plateau which includes the Grand Canyon is also the result of broad tectonic uplift followed by river erosion.

Isostatic Uplift

The removal of mass from a region will be isostatically compensated by crustal rebound. If we take into consideration typical crustal and mantle densities, erosion of an average 100 meters of rock across a broad, uniform surface will cause the crust to isostatically rebound about 85 meters and will cause only a 15-meter loss of mean surface elevation. An example of isostatic uplift would be post-glacial rebound following the melting of continental glaciers and ice sheets. The Hudson Bayregion of Canada, the Great Lakes of Canada and the United States, and Fennoscandia are currently undergoing gradual rebound as a result of the melting of ice sheets 10,000 years ago.

Crustal thickening, which for example is currently occurring in the Himalaya due to the continental collision between the Indian and the Eurasian plates, can also lead to surface uplift; but due to the isostatic sinking of thickened crust, the magnitude of surface uplift will only be about one-sixth of the amount of crustal thickening. Therefore, in most convergent settings isostatic uplift plays a relatively small role and high peak formation can be more attributed to tectonic processes.Direct measures of the elevation change of the land surface can only be used to estimate erosion or bedrock uplift rates when other controls (such as changes in mean surface elevation, volume of eroded material, timescales and lags of isostatic response, variations in crustal density) are known.

Coral Islands

In a few cases, tectonic uplift can be seen in the cases of coral islands. This is evidenced by the presence of various oceanic islands composed entirely of coral, which otherwise appear to be high islands (*i.e.*, islands of volcanic origin). Examples of such islands are found in the Pacific, notably the three phosphate islets, Nauru, Makatea, and Banaba as well as Maré and Lifou in New-Caledonia, Fatu Huku in the Marquesas Islands and Henderson Island in the Pitcairn Islands. The uplift of these islands is the result of the movement of oceanic tectonic plates. Sunken islands or guyots with their coral reefs are the result of crustal subsidence as the oceanic plate carries the islands to deeper or lower oceanic crust areas.

Uplift vs. Exhumation

The word "uplift" refers to displacement contrary to the direction of the gravity vector, and displacement is only defined when the object being displaced and the frame of reference is specified. Molnar and England, identify three kinds of displacement to which the term "uplift" is applied:

1. Displacement of the Earth's surface with respect to the geoid. This is what we refer to as "surface uplift"; and surface uplift can be defined by averaging elevation and changes in elevation over surface areas of a specified size.
2. The "uplift of rocks" refers to the displacement of rocks with respect to the geoid.
3. The displacement of rocks with respect to the surface is called exhumation.

This simple equation relates the three kinds of displacement:

Surface uplift = uplift of rock - exhumation

Measuring uplift and exhumation can be tricky. Measuring the uplift of a point requires measuring its elevation change – usually geoscientists are not trying to determine the uplift of a singular point, but rather the uplift over a specified area. Accordingly, the change in elevation of all points on the surface of that area must be measured, and the rate of erosion must be zero or minimal. Also, sequences of rocks deposited during that uplift must be preserved. Needless to say, in mountain ranges where elevations are far above sea level these criteria are not always easily met. Paleoclimatic restorations though can be very valuable; these studies involve inferring changes in climate in an area of interest from changes with time of flora/fauna that is known to be sensitive to temperature and rainfall. The magnitude of the exhumation a rock has been subjected to may be inferred from geobarometry (measuring previous pressure and temperature history of a rock or assemblage). Knowing the pressure and temperature history of a region can yield an estimate of the ambient geothermal gradient and bounds on the exhumation process; however, geobarometric/geothermometric studies do not produce a rate of exhumation (or any other information on time). One can infer exhumation rates from fission tracks and from radiometric agesas long as one has an estimated thermal profile.

Weathering

Weathering refers to the response of rocks to the physical and chemical environment at the interface of the atmosphere and the land surface. Rocks exposed for any considerable period of time to the action of the atmosphere are generally converted into soils by a process called weathering.

Weathering is the alteration that rocks undergo by their exposure to air, water and organic matter. The collective processes of rock disintegration and decay are termed weathering. The surface of the earth's crust is continuously exposed to the changing weather of the atmosphere. The environment becomes different from the one when most rocks and minerals were formed.

These materials as they adjust themselves to the changed environment undergo changes. Thus the crust of the earth or bedrock which is solid and resistant at depth is usually cracked and jointed near the surface where the whole mass may be overlain or covered by loose rock fragments of various sizes which compose the regolith.

It is a matter of simple observation that rocks exposed at the earth's surface show the effects of attack by the agents of weather. No rock can endure long the weather attacks. Some rocks soften and crumble, others dissolve and are carried away by water. In the weathering process new secondary matter may be produced that may later form sedimentary rocks.

Moisture-laden atmosphere penetrates the pores and cracks of the bedrock and this causes changes which lead to the formation of most of the regolith. Thus the crust is continuously undergoing changes.

The rocks present at or near the surface are subjected to conditions that ultimately change their physical and chemical composition. Since the chemical and mechanical factors producing such effect are associated with weather, these processes are called weathering.

It is a matter of observation that century old stone structures present their worn out and decayed state unless protected and repaired annually. Monuments, roofing slates, roads, foundation, concrete buildings, steel bridges etc. are all subject to weathering.

Man no doubt does his best to resist the action of weathering, by using resistant materials for the structures he builds and paints the exposed surfaces, but these are all his efforts only to delay the disintegrations and decompositions.

Rocks are attacked because they directly come into contact with air and water i.e. the weathering agents. Thus, weathering is related to the surfaces of the lithosphere, where rocks, air and water come together. Water soaks into the rocks, dissolves and alters minerals and expands by freezing and enlarges joints and fractures.

Weathering includes both physical disintegration and chemical decomposition. In general the former causes rocks to crumble or break into fragments and the latter cause them to decay. Although quite different, these processes act simultaneously and it may not always be possible to assign specific results to one or the other. The physical phase of weathering may dominate in some circumstances and chemical phase in others.

There are three processes in which rocks undergo weathering as follows:

1. Mechanical disintegration of a rock mass at the surface due to changes in temperature and also as wind, water, ice and rock fragments carried by them buffet or press against it or forces it apart.

2. Chemical reactions between the original minerals of the rock, the near surface water and the oxygen of the atmosphere to produce new minerals which are stable under the conditions at the earth's surface and remove other soluble constituents.

3. Biological activity which produces organic acids, thus adding to the chemical reactions and which may also be an agent assisting mechanical disintegration.

Mechanical Weathering or Physical Weathering

Mechanical or physical weathering refers to changes involving form only. Due to this type of weathering large solid masses may break into loose fragments varying in size and shape but maintaining their original composition. Such processes that break down rocks without changing their chemical composition are called physical or mechanical weathering.

Mechanical weathering can be of two types viz. block disintegration and granular disintegration. Block disintegration takes place due to development of joints which break the rock mass into a number of smaller individual blocks or fragments. Granular disintegration takes place due to loss of cohesion between individual particles rendering the rock to incoherent granular pieces.

Granular disintegration is limited to the coarse grained rocks and affects particular rocks like coarse textured granites. Block disintegration affects rocks of all textures and is particularly conspicuous in the finer textured varieties. In addition to block and granular disintegration, impact and abrasion also can cause disintegration of rocks.

Physical weathering may be due to the following:

1. Differential thermal expansion,
2. Temperature variations,
3. Undermining,
4. Abrasion, grinding and impact,
5. Exfoliation,
6. Frost action,
7. Plant and Animal action,
8. Pressure unloading.

Differential Thermal Expansion

Minerals in a rock have varying thermal expansion coefficients. Due to rise in temperature differential stresses will be set up. This will lead to granular disintegration of minerals and rocks. The dark coloured mineral have a greater rate of heat absorption than light coloured minerals. This may also contribute to stress formation which may lead to cracks.

Temperature Variations

Rocks are subjected to repeated heating and cooling due to diurnal and seasonal changes of temperatures. During the period of intense heat the outer layers of the rock mass expand introducing tensile stresses. This may cause separation parallel to the rock surface. When there is a considerable fall in temperature, the material near the surface contracts more and this results in radial fissures.

Undermining

Erosion by rivers and sea can lead to rock falls and landslides which may cause rock fractures. This is common along sea coasts where removal of clay from below overlying limestone's occurs. Large scale break down may also result from wind erosion of soft beds at lower levels resulting in harder rocks dropping at the foot of the cliffs.

Abrasion, Grinding and Impact

These three operations reduce the size of particles. Abrasion is typical of rubbing action of debris laden ice masses passing over a rock floor. Grinding is the effect produced by small fragments being caught up between larger ones and being ground down to almost rock flour. Such actions are likely along river channels and along coasts. Impact refers to sudden collision of rock bodies which leads to flaking and chipping forming fragments.

Exfoliation

This refers to the scaling off or peeling off successive shells from the rock surface. Exfoliation is observed in coarse grained rocks containing feldspar. When the rock surface gets wet, the moisture penetrates the pores and the crevices between the mineral grains and reacts with feldspar. As a consequence of the chemical reaction, a new substance namely kaolin is formed which is a form of clay.

This clay has a larger volume than the originally present feldspar. This expansion pries loose the surrounding mineral grains. As a result of this action, a thin shell of surface material flakes away. This process is repeated due to successive wettings of the rock surface.

Frost Action

Frost action is due to a contrasting property of water. We know most materials expand when heated and contract when cooled. This is true of water except that when water is cooled down from 4°C to 0 °C it expands.

The extent of expansion is most at 0 °C as it solidifies into ice, the volume increasing by 9 per cent. Such expansion of water as it cools and solidifies can exert enormous forces producing stresses of many thousands of Newton's per square millimetre. When rainwater, melting snow or condensation seeps into any pores or crevices in rocks, as the temperature falls below the freezing point, the water seeping into the cracks and pores, changes into ice.

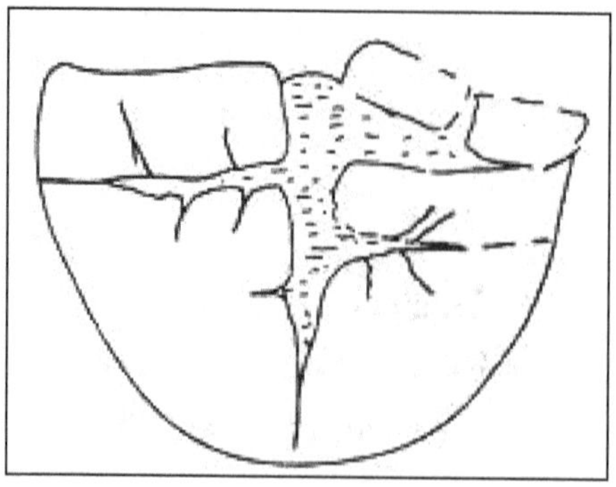

Frost Wedging.

The expanding ice exerts enormous pressure against the confining rock, acting like a wedge and widens and extends the opening. Thereafter when the ice thaws, the water seeps deeper into the opening. As the water refreezes, the process is repeated. Such repeated thawing and freezing of water, i.e. frost action breaks the rock apart.

Frost action is prominent where the bed rock is directly exposed to the atmosphere and where moisture exists and the temperature fluctuates often above and below the freezing point of water.

Such condition exist in winter in temperate climates and they can also occur on mountain tops and also in regions of high elevation in spring or fall. Day time temperatures rise above the freezing point causing snow and ice to melt and as the temperature falls below the freezing point again during the night, producing frost action.

Due to frost action on cliffs, the broken loose fragments fall to the base of the cliff. As this process continues, a pile of fragments called Tallus slope accumulates at the base of the cliff. Pot holes on roads in cold regions are due to frost action on exposed road surfaces.

Plant and Animal Action

Rocks can break down into smaller pieces by the interaction on them by plants and animals. When a rock develops cracks small rock particles and soil are washed into such a crack by rain or wind. If a seed should drop into such a crack, it can germinate and can grow to become a plant.

Such a plant can send its rootlets deeper into the rocks in search of water. As the growing rootlets thicken, they press against the sides of the crack and over a period of time can break the rock. Roots of tiny plants like lichens and mosses produce a rock-dissolving acid as they grow and decay further accelerating breaking down of the rocks.

Animals (with the exception of humans) also contribute to the weathering of rocks. Earthworms can bring particles to the surface. These particles are exposed to the atmosphere and subjected to further break down. Ants, termites, moles and such burrowing animals can cause weathering. The burrows created by them permit air and water to penetrate to cause weathering of the underlying rock.

Rock quarrying, strip mining are examples of human activities where rocks are broken. In addition, such activities expose huge quantities of fresh rock to other weathering processes.

Pressure Unloading

Rocks formed at great depths are under high pressure. Very high compressive stresses are developed in them which cannot be released because of pressure.

Certain forces within the earth bring these rocks to the surface and in such situations the pressure is released leading to expansion and release of stress. In this process the rocks develop large cracks or joints where they are weak. Unloading can also take place when very heavy glaciers melt away and the pressure is released. Physical weathering exposes large surface areas necessary for chemical activity to take place.

Chemical Weathering of Rocks

Chemical weathering is a process in which rocks are broken down changing their chemical compositions. Most of the rocks are formed in an environment very much different from the environment prevailing at the surface of the earth. Many of the substances present in the atmosphere are not present in the environment where the rocks are formed.

Hence, when the mineral of the rock are exposed to the substance of the atmosphere, chemical reactions take place resulting in the formation of new compounds, whose properties differ from those of the original minerals. These changes weaken the structure of the rock and as a consequence, the rock gets broken by physical weathering.

It is worthy to note the following general characteristics of chemical reactions in the context of different weathering environments.

1. Chemical reactions tend to proceed more rapidly at higher temperatures.
2. For efficient reaction, the reactants must be brought together quickly and easily and the products must be removed. In nature water generally supplies the reactants to the mineral surfaces and flushes away the reaction products.
3. The smaller the reacting grains, the more rapidly the chemical reactions proceed to completion. All the above factors play a role in the chemical weathering process. Local climate controls the average temperature of the reaction and the supply of water for reaction.

Grain size of the mineral reactants depends largely on the process of mechanical weathering (disintegration) of rocks as well as abrasion and breakage during transportation. The length of time available for weathering reactions depends on the rate of erosion and therefore the rate of uplift or subsidence.

If erosion or deposition occurs rapidly, then weathering reactions will be interrupted because sediments will be buried and removed from the weathering environment; if erosion or deposition occurs slowly, then weathering reactions can proceed for a longer time. Chemical weathering is brought mainly by oxygen, carbon dioxide and water.

Oxidation

Oxidation means the combination of oxygen with other substances. This is an important chemical weathering process. Most of the minerals containing iron like Magnetite, Pyrite Amphibole. Biotites are readily affected by oxygen, of which Haematite (Fe_2O_3) and Magnetite (Fe_3O_4) are very common.

Presence of water during oxidation can result in another reaction. A compound of Iron, oxygen and water called goethite can be formed. Goethite is yellowish brown in colour. When goethite is dehydrated, hematite is formed. Presence of hematite or goethite in soils gives a reddish or yellow brown colour.

Oxidation of iron oxide in the presence of water:

$$\underset{\text{iron oxide}}{4FeO} + \underset{\text{Water}}{2H_2O} + \underset{\text{Oxygen}}{O_2} \rightarrow \underset{\text{goethite}}{4(FeO.OH)}$$

Dehydration of goethite:

$$\underset{\text{goethite}}{2[FeO.OH]} \rightarrow \underset{\text{hematite}}{Fe_2O_3} + \underset{\text{Water}}{2H_2O}$$

Oxidation causes breaking down of the rock due to the following effect. When oxygen combines with iron, the chemical bonds between iron and other elements are broken thus weakening the structure. Even Aluminium and Silicon when subjected to oxidation forming oxides can get weakened in structure.

Hydration, Hydrolysis and Solution

Water present on the earth's surface is an important agent of chemical weathering. A reaction of water with another substance is called hydration.

Ex: Hydration of anhydrite to form gypsum,

$$\underset{\text{Anhydrite}}{CaSO_4} + \underset{\text{Water}}{2H_2O} \rightarrow \underset{\text{gypsum}}{CaSO_4.2H_2O}$$

Water can also break into hydrogen ions (H^+) and hydroxide ions (OH^-). If these ions replace the ions of the mineral, the reaction is called hydrolysis. Common minerals that undergo hydrolysis are feldspar, amphibole and biotite. This process results in swelling and crumbling to powder.

Water can dissolve rock matter and bring about weathering. This process is called weathering by solution. Halite (Rock salt) and gypsum are examples of minerals soluble in water. As water slowly dissolves some minerals out of the rock, the surrounding rock minerals are exposed for further weathering.

In some cases the structure of the rock can get weakened due to vacant cavities created leading to the crumbling of the rock. The minerals dissolved in solution can chemically react with each other to form new compounds. If the resulting compounds are insoluble in water, they may precipitate out.

Carbonation

Chemical combination of carbon dioxide with another substance is called carbonation. Carbon dioxide in gas state may have no effect on rocks. But, when carbon dioxide comes into contact with water, carbonic acid is formed which can act on common rock minerals. Minerals containing sodium, potassium, magnesium and calcium are affected by carbonic acid to form carbonates.

The mineral calcite is severely affected by carbonic acid to almost destruction. Limestone is totally dissolved by carbonic acid present in groundwater or rainwater. As groundwater containing carbonic acid seeps through bed rocks consisting of calcite, spectacular caverns are formed, due to the formation of very large holes.

Other Chemical Factor

There are also other acids besides carbonic acid that attack rocks and minerals. Some of these acids are produced during the decay of organic matter. Some acids are produced in the form of waste products of certain plants and animals. These acids get dissolved in rainwater and seep through the soil reaching the bedrock and chemically act with the rock.

Some primitive plants like lichens can grow on bare rock when the rock is wet and lie dormant when the rock is dry, secretion from lichens corrode the rock surface dissolving out the mineral nutrients loosening the mineral particles. The loosened mineral particles along with dust accumulate in the crevices of the rock. Some seeds can get into these soil particles and grow leading to further physical weathering.

Human activities also become sources of acids which can cause rock weathering. Homes, cars, buses, trucks etc. release large quantities of waste gases and other pollutants into the atmosphere. Many of these like oxides of nitrogen and sulphur chemically react with water to form reactive acids.

Bacteria can also exercise an important influence in promoting rock disintegration and decomposition. Some of them are known to give out nitric acid which can chemically act on the rocks. The microscopic bacteria penetrate into every little crevice produced by atmospheric agencies and throughout the long periods of time bring about disintegration of surface rocks, their period of activity being limited to the summer months.

They have been noticed on rocks of widely different character like, granites, schists, limestone's, sandstones, volcanic rocks and on high mountain peaks as well as lower levels. It is also reported that certain variety of ants continuously pour out carbonic acid into the ground leading to weathering.

Certain other species of ants known as saubas or sauvas live in big colonies, burrowing in the earth, where they excavate chambers with galleries radiating in every direction into which they carry large quantities of leaves.

In regions of industrial complexes acids are in alarming quantity. Water in rainfall in these regions contains considerable quantity of acid and the rain is often called acid rain. Rocks can get weathered and broken down by the action of acid rain. Acid rain can also breakdown man-made structures and can damage plant and animal life.

Forms of Chemical Weathering

Solution Weathering

This is another form of chemical weathering. This occurs when minerals dissolve in water (go in solution). This happens because some types of rock are easily dissolved in rainwater. Weathering by solution typically produces rather smooth scalloped surfaces. For example, soft calcite and gypsum often show evidence of solution weathering.

Spheroidal Weathering

Spheroidal weathering refers to the alteration of joint blocks of rock progressively inward from their rims. The edge region of the fractures of the rock is partially or wholly converted into clay or other products.

While the inner regions of the rock remain relatively fresh and solid, the exteriors are subjected to differential expansions and the material in this region gets loosened along concentric joints. The cores formed range in size from boulder to pebbles. They get rounded by weathering. This type of weathering is due to the rock subjected to both mechanical and chemical weathering.

First the rocks are split developing joints. The separated blocks of the rock undergo chemical weathering due to which the edges and the surfaces of the individual separated blocks undergo corrosion. As a consequence, the separated blocks are changed into rounded boulders.

Differential Weathering

We often observe in many road cut/out crops layer of rock that are all weathering at different rates making the out crop look like an uneven stack of flat rocks. This is called differential weathering.

This occurs when the layers in an outcrop contain more than one type of rock for example, certain ancient marine environments may deposit separate layers of sand and silt creating an out crop of sandstone and shale. When these two types of rock weather, the result is often differential weathering, in which the sandstones are more resistant to weathering than the shales.

Diagenesis

A form of diagenesis is permineralization, in which buried organisms are replaced by minerals. These trilobites (Lloydolithus) were replaced by pyrite during a specific type of permineralization called pyritization.

Diagenesis is the change of sediments or existing sedimentary rocks into a different sedimentary rock during and after rock formation (lithification), at temperatures and pressures less than that required for the formation of metamorphic rocks. It does not include changes from weathering. It is any chemical, physical, or biological change undergone by a sediment after its initial deposition, after its lithification. This process excludes surface alteration (weathering) and metamorphism. These changes happen at relatively low temperatures and pressures and result in changes to the rock's original mineralogyand texture. There is no sharp boundary between diagenesis and metamorphism, but the latter occurs at higher temperatures and pressures. Hydrothermal solutions, meteoric groundwater, porosity, permeability, solubility, and time are all influential factors.

After deposition, sediments are compacted as they are buried beneath successive layers of sediment and cemented by minerals that precipitate from solution. Grains of sediment, rock fragments and fossils can be replaced by other minerals during diagenesis. Porosity usually decreases during diagenesis, except in rare cases such as dissolution of minerals and dolomitization.

The study of diagenesis in rocks is used to understand the geologic history they have undergone and the nature and type of fluids that have circulated through them. From a commercial standpoint, such studies aid in assessing the likelihood of finding various economically viable mineral and hydrocarbon deposits.

The process of diagenesis is also important in the decomposition of bone tissue.

Role of Diagenesis in Anthropology and Paleontology

Originally calcitic crinoid stem (in cross-section) diagenetically replaced by marcasite in a siderite concretion; Lower Carboniferous.

The term diagenesis, literally meaning "across generation", is extensively used in geology. However, this term has filtered into the field of anthropology, archaeologyand paleontology to describe the changes and alterations that take place on skeletal (biological) material. Specifically, diagenesis "is the cumulative physical, chemical and biological environment; these processes will modify an organic object's original chemical and/or structural properties and will govern its ultimate fate, in terms of preservation or destruction". In order to assess the potential impact of diagenesis on archaeological or fossil bones, many factors need to be assessed, beginning with elemental and mineralogical composition of bone and enveloping soil, as well as the local burial environment (geology, climatology, groundwater).

The composite nature of bone, comprising one-third organic (mainly proteincollagen) and two thirds mineral (calcium phosphate mostly in the form of hydroxyapatite) renders its diagenesis more complex. Alteration occurs at all scales from molecular loss and substitution, through crystallite reorganization, porosity and microstructural changes, and in many cases, to disintegration of the complete unit. Three general pathways of the diagenesis of bone have been identified:

1. Chemical deterioration of the organic phase.
2. Chemical deterioration of the mineral phase.
3. (Micro) Biological attack of the composite.

They are as follows:

1. The dissolution of collagen depends on time, temperature and environmental pH. At high temperatures, the rate of collagen loss will be accelerated and extreme pH can cause collagen swelling and accelerated hydrolysis. Due to the increase in porosity of bones through collagen loss, the bone becomes susceptible to hydrolytic infiltration where the hydroxyapatite, with its affinity for amino acids, permits charged species of endogenous and exogenous origin to take up residence.
2. The hydrolytic activity plays a key role in the mineral phase transformations that exposes the collagen to accelerated chemical- and bio-degradation. Chemical changes affect crystallinity. Mechanisms of chemical change, such as the uptake of F^- or CO_3^- may cause recrystallization where hydroxyapatite is dissolved and re-precipitated allowing for the incorporation of substitution of exogenous material.
3. Once an individual has been interred, microbial attack, the most common mechanism of bone deterioration, occurs rapidly. During this phase, most bone collagen is lost and porosity is increased. The dissolution of the mineral phase caused by low pH permits access to the collagen by extracellular microbial enzymes thus microbial attack.

Role of Diagenesis in Hydrocarbon Generation

When animal or plant matter is buried during sedimentation, the constituent organic molecules (lipids, proteins, carbohydrates and lignin-humic compounds) break down due to the increase in temperature and pressure. This transformation occurs in the first few hundred meters of burial and results in the creation of two primary products: kerogensand bitumens.

It is generally accepted that hydrocarbons are formed by the thermal alteration of these kerogens (the *biogenic* theory). In this way, given certain conditions (which are largely temperature-dependent) kerogens will break down to form hydrocarbons through a chemical process known as cracking, or catagenesis.

A kinetic model based on experimental data can capture most of the essential transformation in diagenesis, and a mathematical model in a compacting porous medium to model the dissolution-precipitation mechanism. These models have been intensively studied and applied in real geological applications.

Diagenesis has been divided, based on hydrocarbon and coal genesis into: eodiagenesis (early), mesodiagenesis (middle) and telodiagenesis (late). During the early or eodiagenesis stage shales lose pore water, little to no hydrocarbons are formed and coal varies between lignite and sub-bituminous. During mesodiagenesis, dehydration of clay minerals occurs, the main development of oil genesis occurs and high to low volatile bituminous coals are formed. During telodiagenesis, organic matter undergoes cracking and dry gas is produced; semi-anthracite coals develop.

Early diagenesis in newly formed aquatic sediments is mediated by microorganisms using different electron acceptors as part of their metabolism. Organic matter is mineralized, liberating gaseous carbon dioxide (CO_2) in the porewater, which, depending on the conditions, can diffuse into the water column. The various processes of mineralization in this phase are nitrification and denitrification, manganese oxide reduction, iron hydroxide reduction, sulfate reduction, and fermentation.

Role of Diagenesis in Bone Decomposition

Diagenesis alters the proportions of organic collagen and inorganic components (hydroxyapatite, calcium, magnesium) of bone exposed to environmental conditions, especially moisture. This is accomplished by the exchange of natural bone constituents, deposition in voids or defects, adsorption onto the bone surface and leaching from the bone.

Metamorphism

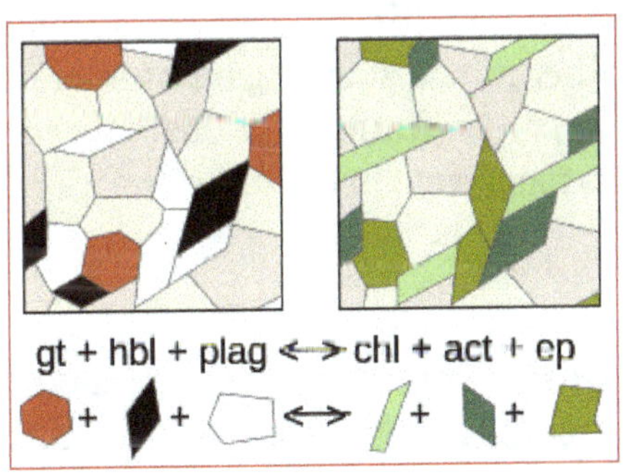

Schematic representation of a metamorphic reaction. Abbreviations of minerals: act = actinolite; chl = chlorite; ep = epidote; gt = garnet; hbl = hornblende; plag = plagioclase. Two minerals represented in the figure do not participate in the reaction, they can be quartz and K-feldspar. This reaction takes place in nature when a mafic rock goes from amphibolite facies to greenschist facies.

A cross-polarized thin section image of a Garnet-Mica-Schist from Salangen, Norway showing the strong strain fabric of schists. The black (isotropic) crystal is garnet, the pink-orange-yellow (commonly referred to as high order) colored strands are muscovite mica, with the brown crystals are biotitemica. The grey and white (low first-order) crystals are quartz and (limited) feldspar.

Metamorphism is the change of minerals or geologic texture (distinct arrangement of minerals) in pre-existing rocks (protoliths), without the protolith melting into liquid magma (a solid-state change). The change occurs primarily due to heat, pressure, and the introduction of chemically active fluids. The chemical components and crystal structures of the minerals making up the rock may change even though the rock remains a solid. Changes at or just beneath Earth's surface due to weathering or diagenesis are not classified as metamorphism.Metamorphism typically occurs between diagenesis (maximum 200 °C), and melting (~850 °C).

The geologists who study metamorphism are known as "metamorphic petrologists." To determine the processes underlying metamorphism, they rely heavily on statistical mechanics and experimental petrology.

Three types of metamorphism exist: contact, dynamic, and regional. Metamorphism produced with increasing pressure and temperature conditions is known as *prograde metamorphism*. Conversely, decreasing temperatures and pressure characterize *retrograde metamorphism*.

Solid State Recrystallization and Neocrystalization

Metamorphic rocks can change without melting. Heat causes atomic bonds to break, and the atoms move and form new bonds with other atoms, creating new minerals with different chemical components or crystalline structures (neocrystallization), or enabling recrystallization. When pressure is applied, somewhat flattened grains that orient in the same direction have a more stable configuration.

Limits

The temperature lower limit on what is considered to be a metamorphic process is generally considered to be 100 – 200 °C; this excludes diagenetic changes due to compaction and the formation of sedimentary rocks.

The upper boundary of metamorphic conditions is related to the onset of melting processes in the rock. The maximum temperature for metamorphism is typically 700 – 900 °C, depending on the pressure and on the composition of the rock. Migmatites are rocks formed at this upper limit, which contains pods and veins of material that has started to melt but has not fully segregated from the refractory residue. Since the 1980s it has been recognized that rocks are rarely dry enough and of a refractory enough composition to record without melting "ultra-high" metamorphic temperatures of 900 – 1100 °C. The metamorphic process usually has to be over pressure of at least 100 mega pascals but below 300 mega pascals, the depth of 100 mega pascals varies depending on what type of rock is applying pressure.

Types

Regiona

Regional or *Barrovian* metamorphism covers large areas of continental crust typically associated with mountain ranges, particularly those associated with convergent tectonic plates or the roots of previously eroded mountains. Conditions producing widespread regionally metamorphosed rocks occur during an orogenic event. The collision of two continental plates or island arcs with continental plates produce the extreme compressional forces required for the metamorphic changes typical of regional metamorphism. These orogenic mountains are later eroded, exposing the intensely deformed rocks typical of their cores. The conditions within the subducting slab as it plunges toward the mantle in a subduction zone also produce regional metamorphic effects, characterized by paired metamorphic belts. The techniques of structural geologyare used to unravel the collisional history and determine the forces involved. Regional metamorphism can be described and classified into metamorphic facies or metamorphic zones of temperature/pressure conditions throughout the orogenic terrane.

Contact Thermal

A metamorphic aureole in the Henry Mountains, Utah. The greyish rock on top is the igneous intrusion, consisting of porphyritic granodiorite from the Henry Mountains laccolith, and the pinkish rock on the bottom is the sedimentary country rock, a siltstone. In between, the metamorphosed siltstone is visible as both the dark layer (~5 cm thick) and the pale layer below it.

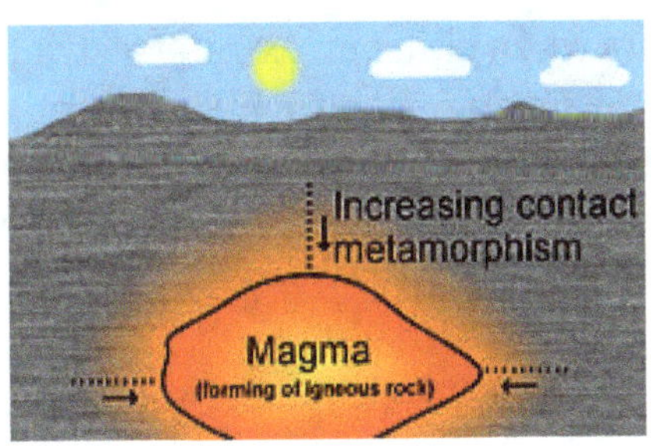

Contact metamorphism occurs typically around intrusive igneous rocks as a result of the temperature increase caused by the intrusion of magma into cooler country rock. The area surrounding the intrusion where the contact metamorphism effects are present is called the *metamorphic aureole*. Contact metamorphic rocks are usually known as hornfels. Rocks formed by contact metamorphism may not present signs of strong deformation and are often fine-grained.

Contact metamorphism is greater adjacent to the intrusion and dissipates with distance from the contact. The size of the aureole depends on the heat of the intrusion, its size, and the temperature difference with the wall rocks. Dikesgenerally have small aureoles with minimal metamorphism whereas large ultramafic intrusions can have significantly thick and well-developed contact metamorphism.

The metamorphic grade of an aureole is measured by the peak metamorphic mineral which forms in the aureole. This is usually related to the metamorphic temperatures of pelitic or aluminosilicate rocks and the minerals they form.The metamorphic grades of aureoles are andalusite hornfels, sillimanite hornfels, pyroxene hornfels.

Magmatic fluids coming from the intrusive rock may also take part in the metamorphic reactions. An extensive addition of magmatic fluids can significantly modify the chemistry of the affected rocks. In this case the metamorphism grades into metasomatism. If the intruded rock is rich in carbonate the result is a skarn. Fluorine-rich magmatic waters which leave a cooling granite may often form greisens within and adjacent to the contact of the granite. Metasomatic altered aureoles can localize the deposition of metallic ore minerals and thus are of economic interest.

A special type of contact metamorphism, associated with fossil fuel fires, is known as pyrometamorphism.

Hydrothermal

Hydrothermal metamorphism is the result of the interaction of a rock with a high-temperature fluid of variable composition. The difference in composition between an existing rock and the invading fluid triggers a set of metamorphic and metasomaticreactions. The hydrothermal fluid may be magmatic (originate in an intruding magma), circulating groundwater, or ocean water. Convective circulation of hydrothermal fluids in the ocean floor basalts produces extensive hydrothermal metamorphism adjacent to spreading centers and other submarine volcanic areas. The fluids eventually escape through vents on the ocean floor known as black smokers. The patterns of this hydrothermal alteration are used as a guide in the search for deposits of valuable metal ores.

Shock

This kind of metamorphism occurs when either an extraterrestrial object (a meteorite for instance) collides with the Earth's surface or during an extremely violent volcanic eruption. Impact metamorphism is, therefore, characterized by ultrahigh pressure conditions and low temperature. The resulting minerals (such as SiO_2 polymorphs coesite and stishovite) and textures are characteristic of these conditions.

Dynamic

Dynamic metamorphism is associated with zones of high to moderate strain such as fault zones. Cataclasis, crushing and grinding of rocks into angular fragments, occurs in dynamic metamorphic zones, giving cataclastic texture.

The textures of dynamic metamorphic zones are dependent on the depth at which they were formed, as the temperature and confining pressure determine the deformation mechanisms which predominate. Within depths less than 5 km, dynamic metamorphism is not often produced because the confining pressure is too low to produce frictional heat. Instead, a zone of breccia or cataclasite is formed, with the rock milled and broken into random fragments. This generally forms a mélange. At depth, the angular breccias transit into a ductile shear texture and into mylonite zones.

Within the depth range of 5–10 km pseudotachylite is formed, as the confining pressure is enough to prevent brecciation and milling and thus energy is focused on discrete fault planes. Frictional heating, in this case, may melt the rock to form pseudotachylite glass.

Within the depth range of 10–20 km, deformation is governed by ductile deformation conditions and hence frictional heating is dispersed throughout shear zones, resulting in a weaker thermal imprint and distributed deformation. Here, deformation forms mylonite, with dynamothermal metamorphism observed rarely as the growth of porphyroblasts in mylonite zones.

Overthrusting may juxtapose hot lower crustal rocks against cooler mid and upper crust blocks, resulting in conductive heat transfer and localised contact metamorphism of the cooler blocks adjacent to the hotter blocks, and often retrograde metamorphism in the hotter blocks. The metamorphic assemblages in this case are diagnostic of the depth and temperature and the throw of the fault and can also be dated to give an age of the thrusting.

Classification of Metamorphic Rocks

Metamorphic rocks are classified by their mineral composition, the source rock, also known as a protolith, and the context (pressure, temperature, hydrological features, etc.) of its formation.

Metamorphic Facies

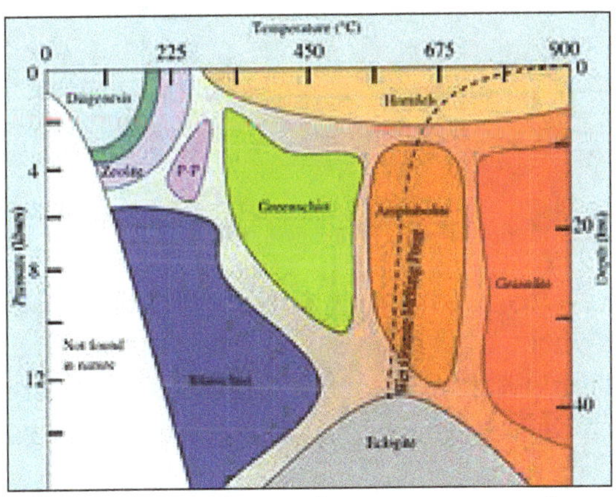

Metamorphic facies with regard to temperature and pressure.

Metamorphic facies are recognizable terranes or zones with an assemblage of key minerals that were in equilibrium under specific range of temperature and pressure during a metamorphic event. The facies are named after the metamorphic rock formed under those facies conditions from basalt. Facies relationships were first described by Pentti Eskola in 1921.

Temperature	Pressure	Facies
Low	Low	Zeolite
Lower Moderate	Lower Moderate	Prehnite-Pumpellyite
Moderate to High	Low	Hornfels
Low to Moderate	Moderate to High	Blueschist
Moderate → High	Moderate	Greenschist→Amphibolite→Granulite
Moderate to High	High	Eclogite

Metamorphic Grades

Metamorphic grade is an informal indication of the amount or degree of metamorphism. A more complete indication of this intensity or degree is provided with the concept of metamorphic facies.

In the Barrovian sequence (described by George Barrow in zones of progressive metamorphism in Scotland), metamorphic grades are also classified by mineral assemblage based on the appearance of key minerals in rocks of pelitic (shaly, aluminous) origin:

Low grade ------------------- Intermediate -------------------- High grade

Greenschist ------------- Amphibolite ---------------------- Granulite

Slate --- Phyllite ---------- Schist --------------------- Gneiss --- Migmatite

Chlorite zone

Biotite zone

Garnet zone

Staurolite zone

Kyanite zone

Sillimanite zone

Metamorphic Processes

Recrystallization

During recrystallization, the grains making up the protolith change shape and size. The identity of the mineral does not change during this process, only the texture. Recrystallization occurs due to heating of the protolith. The temperature at which this occurs can vary depending on the minerals present. Recrystallization generally begins when temperatures reach above half the melting point of the mineral on the Kelvin scale.

(Left) Randomly-orientated grains in a rock before metamorphism.
(Right) Grains align orthogonal to the applied stress if a rock is subjected to stress during metamorphism.

Phase Change

Phase change metamorphism is the creating of new minerals with the same chemical formula as the protolith. This involves a rearrangement of the atoms in the crystals.

Neocrystallization

Neocrystallization involves the creation of new mineral crystals different from the protolith. Chemical reactions digest the minerals of the protolith which yields new minerals. This is a very slow process as it can also involve the diffusion of atoms through solid crystals.

Pressure Solution

Pressure solution is a metamorphic process that requires a rock to be under strong pressure from one direction and in the presence of hot water. During this process mineral of the protolith partially dissolve, diffuse through the water and precipitate elsewhere.

Plastic Deformation

In plastic deformation pressure is applied to the protolith, which causes it to shear or bend, but not break. In order for this to happen temperatures must be high enough that brittle fractures do not occur, but not so high that diffusion of crystals takes place.

Prograde and Retrograde

Metamorphism is further divided into prograde and retrograde metamorphism. Prograde metamorphism involves the change of mineral assemblages (paragenesis) with increasing temperature and (usually) pressure conditions. These are solid state dehydration reactions, and involve the loss of volatiles such as water or carbon dioxide. Prograde metamorphism results in rock characteristic of the maximum pressure and temperature experienced. Metamorphic rocks usually do not undergo further change when they are brought back to the surface.

Retrograde metamorphism involves the reconstitution of a rock via revolatisation under decreasing temperatures (and usually pressures), allowing the mineral assemblages formed in prograde metamorphism to revert to those more stable at less extreme conditions. This is a relatively uncommon process, because volatiles must be present.

Commonly used Diagrams

Petrogenetic Grids

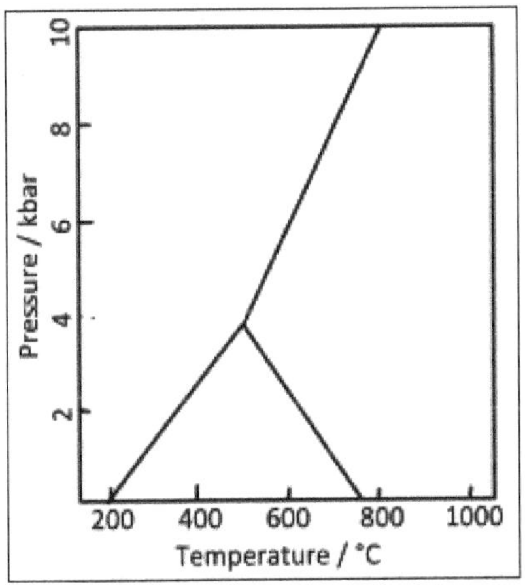

Petrogenetic grid of Al_2SiO_5 (nesosilicates).

A petrogenetic grid is a geologic phase diagram that plots experimentally derived metamorphic reactions at their pressure and temperature conditions for a given rock composition. This allows geologists (specifically: metamorphic petrologists) to determine the pressure and temperature conditions under which rocks metamorphose. The Al_2SiO_5 nesosilicate phase diagram shown is a very simple petrogenetic grid for rocks that only have a composition consisting of aluminum (Al), silicon (Si), and oxygen (O). As the rock undergoes different temperatures and pressure, it could be any of the three given polymorphic minerals. For a rocks that contain more elements than that, the petrogenetic grid quickly becomes more complicated.

Compatibility Diagrams

Compatibility diagrams provide an excellent way to analyze how variations in the rock's composition affect the mineral paragenesis that develops in a rock at particular pressure and temperature conditions.

Mass Wasting

Mass wasting, also known as slope movement or mass movement, is the geomorphic process by which soil, sand, regolith, and rock move downslope typically as a solid, continuous or discontinuous mass, largely under the force of gravity, but frequently with characteristics of a flow as in debris flows and mudflows. Types of mass wasting include creep, slides, flows, topples, and falls, each with its own characteristic features, and taking place over timescales from seconds to hundreds of years. Mass wasting occurs on both terrestrial and submarine slopes, and has been observed on Earth, Mars, Venus, and Jupiter's moon Io.

When the gravitational force acting on a slope exceeds its resisting force, slope failure (mass wasting) occurs. The slope material's strength and cohesion and the amount of internal friction between material help maintain the slope's stability and are known collectively as the slope's shear strength. The steepest angle that a cohesionless slope can maintain without losing its stability is known as its angle of repose. When a slope made of loose material possesses this angle, its shear strength perfectly counterbalances the force of gravity acting upon it

Talus cones produced by mass wasting, north shore of Isfjord, Svalbard, Norway.

Mass wasting may occur at a very slow rate, particularly in areas that are very dry or those areas that receive sufficient rainfall such that vegetation has stabilized the surface. It may also occur at very high speed, such as in rockslides or landslides, with disastrous consequences, both immediate and delayed, e.g., resulting from the formation of landslide dams.

Mass wasting at Palo Duro Canyon, West Texas.

A rockfall in Grand Canyon National Park.

Factors that change the potential of mass wasting include: change in slope angle, weakening of material by weathering, increased water content; changes in vegetation cover, and overloading.

Volcano flanks can become over-steep resulting in instability and mass wasting. This is now a recognised part of the growth of all active volcanoes. It is seen on submarine as well as surface volcanoes: Loihi in the Hawaiian volcanic chain and Kick ‹em Jenny in the Caribbean volcanic arc are two submarine volcanoes that are known to undergo mass wasting. The failure of the northern flank of Mount St Helens in 1980 showed how rapidly volcanic flanks can deform and fail.

Role of Water

Water can increase or decrease the stability of a slope depending on the amount present. Small amounts of water can strengthen soils because the surface tension of water increases soil cohesion. This allows the soil to resist erosion better than if it were dry. If too much water is present the water may act to increase the pore pressure, reducing friction, and accelerating the erosion process and resulting in different types of mass wasting (i.e. mudflows, landslides, etc.). A good example of this is to think of a sand castle. Water must be mixed with sand in order for the castle to keep its shape. If too much water is added the sand washes away, if not enough water is added the sand falls and cannot keep its shape. Water also increases the mass of the soil, this is important because an increase in mass means that there will be an increase in velocity if mass wasting is triggered. Saturated water, however, eases the process of mass wasting in that the rock and soil debris are easily washed down-slope.

Types

Based on how the soil, regolith or rock moves downslope as a whole, mass movements can be broadly classified as creeps and landslides:

Creep

Soil creep is a slow and long term mass movement. The combination of small movements of soil or rock in different directions over time are directed by gravity gradually downslope. The steeper the slope, the faster the creep. The creep makes trees and shrubs curve to maintain their perpendicularity, and they can trigger landslides if they lose their root footing. The surface soil can migrate under the influence of cycles of freezing and thawing, or hot and cold temperatures, inching its way towards the bottom of the slope forming terracettes. Landslides are often preceded by soil creep accompanied with soil sloughing – loose soil that falls and accumulates at the base of the steepest creep sections.

Landslide

A landslide, also called a landslip, is a slow or rapid movement of a large mass of earth and rocks down a hill or a mountainside. Little or no flowage of the materials occurs on a given slope until heavy rain and resultant lubrication by the same rainwater facilitate the movement of the materials, causing a landslide to occur.

In particular, if the main feature of the movement is a slide along a planar or curved surface, the landslide is termed slump, earth slide, debris slide or rock slide, depending on the prevailing material.

Movement of soil and regolith that more resembles fluid behavior is called a flow. These include avalanches, mudflows, debris flows, earth flow, lahars and sturzstroms. Water, air and ice are often involved in enabling fluid-like motion of the material.

A fall, including rockfall and debris fall, occurs where regolith cascades down a slope, but is not of sufficient volume or viscosity to behave as a flow. Falls are promoted in rocks which are characterized by the presence of vertical cracks. Falls can also result from undercutting by running water as well as by waves. They usually occur at very steep slopes such as a cliff face. The rock material may be loosened by earthquakes, rain, plant-root wedging, and expanding ice, among other things. The accumulation of rock material that has fallen and resides at the base of the structure is known as talus.

Triggers

Soil and regolith remain on a hillslope only while the gravitational forces are unable to overcome the frictional forces keeping the material in place. Some factors that reduce the frictional resistance relative to the downslope forces, and thus can trigger slope movement, can include:

- Earthquakes.
- Increased overburden from structures.
- Increased soil moisture.
- Reduction of roots holding the soil to bedrock.
- Undercutting of the slope by excavation or erosion.
- Weathering by frost heave or chemical dissolution.
- Bioturbation.

Mitigation

- Afforestation.
- Re-Afforestation.
- Terracing steps on slopes or, more generally, re-modeling its shape.
- Slope stabilization.

References

- Geomorphology, earth-and-environment-geology-and-oceanography-geology-and-oceanography: encyclopedia.com, Retrieved 16 June, 2019
- Weathering-of-rocks-geology- 91151: yourarticlelibrary.com, Retrieved 26 January, 2019
- Weathering-of-rocks-physical-and-chemical-geology/91161: yourarticlelibrary.com, Retrieved 15 March, 2019
- Naylor, Larissa (2005). "The Contributions of Biogeomorphology to the Emerging Field of Geobiology". Palaeogeography, Palaeoclimatology, Palaeoecology. 219.1 (1–2): 35–51. doi:10.1016/j.palaeo.2004.10.013
- Charlton, Ro (2008). Fundamentals of fluvial geomorphology. London: Rutledge. p. 234. ISBN 978-0-415-33454-9
- Burbank, Douglas West., and Robert S. Anderson. Tectonic Geomorphology. Malden, MA: Blackwell Science, 2000. ISBN 978-0632043866

Field Methods in Geology 6

- **Geologic Time Scale**
- **Geologic Mapping**
- **Radiometric Dating**
- **Geophysical Survey**

There are various field methods that are used in geology. Some of these are geologic mapping, radiometric dating and geophysical survey. The geological map is a special purpose map that is made to show features of Earth. The diverse applications of these field methods have been thoroughly discussed in this chapter.

Geology is, at heart, a field science. Even though much work is done in the laboratory and at the computer, geological samples and information must initially be obtained from the context in which they occur in nature. This natural setting would typically be a field locale chosen by the investigating geologist, or by his employers.

The main field instruments used by geologists include the Brunton compass (and/or Silva compass), tape measures, and plane table and alidade. The Brunton compass is a compact device that permits compass bearings to be made upon linear features (including strike lines) and lines connecting any two points. The Brunton may also function as a protractor (when placed upon a map) and as a device for measuring structural dip and vertical angles (using its internal clinometer). The Silva compass is somewhat similar, except that it does not have a bubble level or adjustable clinometer, so a task like measuring a vertical angle is not possible, and strike and dip measurements may not be as accurately made as with a Brunton. This Silva does not come in a rugged case, as does the Brunton, but its design as a flat, blade shape allows it to be used for map work more easily than a Brunton. The geologist's tape measure is usually the reel-in variety, which is marked in meters and feet. A typical tape length is 100 ft (30.5 m). The plane table and alidade are surveying devices used to measure distance and relative elevation. The plane table sits atop a tripod and a geological or topographic map under construction would be taped to its top. The alidade is a telescopic device that can be moved over the map surface as sightings are made. This device allows measure of horizontal distance and elevation. Horizontal distance and elevation of a point on the earth's surface is obtained by viewing through the alidade sight, a rod with a printed scale upon it (called a stadia). Data recorded during this observation are used for recording distance and elevation of the surveyed point upon a map.

The field instruments above would accompany most fully equipped field geologists on an expedition of mapping and sample collection. The geologist would typically also carry along a field notebook, hand lens, hammer, acid bottle, knife, shovels or trowels, sample bags, pens and pencils, aerial photographs and satellite imagery, maps and literature, camping equipment, and a camera. In the modern era, these materials could also be supplemented by a global positioning satellite system (GPS) receiver (for determining location and retracing routes), laptop computers, digital cameras, and portable geophysical equipment (including a gravimeter, altimeter, magnetic susceptibility meter, etc.) Occasionally, a geologist will bring along power tools for cutting or drilling rock or plaster, and burlap for wrapping delicate samples such as fossil bones.

Field methods in geology may be broken down into four main groups: (1) obtaining and marking samples and describing and measuring where they came from in an outcrop; (2) measuring and recording orientation (i.e., altitude) of strata or other planar features; (3) measuring dimensions (height and width); and (4) constructing geologic and topographic maps.

Obtaining and marking samples and describing and measuring where they originate in an outcrop requires observational skills and patience to record all information that might be obtained at one outcrop. Typically, the thickness of strata at an outcrop is recorded in a notebook where the layers are drawn to scale and described as to rock type, grain size, fossil content, color, sedimentary structures, and other attributes. Thickness of strata is measured using a tape measure or a Jacob's staff, which is a long stick made for sighting intervals of equal stratigraphic thickness (usually 5 ft, or 1.5 m). In the field notebook, detail is given about sampling locations and where photographs of the rocks are made. Samples are marked with an arrow indicating 'up' direction and labeled with a number which relates to the notebook number for the outcrop plus a number relating to feet or meters above the base of the stratigraphic section at that location. The same process is followed at each locale. Later, this information is compiled into a measured and described section for each outcrop, which may be used for correlation between outcrops. In terrains where igneous and metamorphic rocks occur, it is usually not so important for sample information to be recorded about the up direction and elevation above the base of outcrop.

Measuring and recording orientation (i.e., attitude) of strata or other planar features is another important field activity that relates to understanding geological structures and to the making of geological maps. Strike, dip direction, and dip magnitude of rock layers and other planar geological features (e.g., foliation) are obtained in as many places as possible within a study area in order to understand completely all the geological structures (i.e., folds and fault patterns) of an area. Analysis of geological structures can help geologists interpret the conditions of deformation of rocks in an area. Generally, the geologist tries to obtain as many orientation or attitude measurements as possible in the field area being studied.

Measuring dimensions (height and width) of an area or of features in an area, is an important aspect of many geological studies. This may be done as an estimate by using the moveable clinometer in a Brunton compass and employing trigonometric relationships to compute the height or width. For example, if one uses a tape measure (or number of foot paces, if the average foot pace of the observer is known) to measure distance to a cliff wall, and then uses the clinometer in his Brunton to measure angle between his eye level and the top of the cliff, a computation of cliff height can be made. In this instance, the cliff height is equal to the person's eye height plus the product of the horizontal distance to the cliff times the tangent of the sighted angle. Geologists sometimes make simple maps, called "pace and compass maps," using the Brunton compass to take bearings and his measured pace length as a distance measure.

Constructing geologic and topographic maps is another field activity that occupies geologists. Geological maps are made by using a base map or set of aerial photographs to record the observed rock type (preferably a measured and described section, as noted above) and rock attitude at numerous locales in the study area. It is the task of the geologist to ultimately fashion a geologic map that is the simplest interpretation of all the surficial data about rock type and rock attitude in the area. Topographic maps are made by plane table and alidade, as noted above, and these may form the base map for geological mapping studies because surficial elevation is important in interpreting physical relationships between rock formations.

Other types of geological field work include reconnaissance studies of areas where detailed mapping is yet to be done, geological sample analysis conducted on-site at drilling operations, geophysical studies (where the objective is to collect data such as gravity strength, magnetic characteristics, etc.), surface- and groundwater studies (where the emphasis is upon water distribution, quality, and its relationship to geologic features), economic geology studies (where mines and excavations are studied and areas explored for the value of potential new mining), engineering geology field work (where studies assess the impact of human disturbance upon rock and soil stability), and many others.

Geologic Time Scale

Geological time scale is a table showing the sequence of geological periods in the history of earth. It also shows the lengths of time different geological periods are assumed to have occupied. It is measured in millions of years. It has been constructed by studying rock strata, where these have been exposed by excavations or mining or where rivers have cut deeply into the earth's crust.

Scientists have proved that lower the rock layer the older it is and the more ancient are the fossils it contains. With a knowledge of rates of erosion and deposition, the intervals occupied by different periods can be estimated roughly by measuring the relative thicknesses of the rock strata.

By the technique of radiometric dating (i.e. by measuring the rates of decay of radioactive materials in the rocks) more accurate dating can now be provided. In the table of geological time scale, the results of all such works are summarized.

Table also shows the types of plants that have been found during different periods. The earth has been estimated to be about 4600 million years old. Life is believed to have originated about 3000-3500 million years ago on the earth.

Major Divisions and Subdivisions of Geological Time Scale

Geological time has been divided into some major divisions called eras. Each era has been divided into periods, which are then subdivided into epochs.

Different Periods are Recognized Mainly on the basis of the following:

- Changes in composition of fossils, and

- The occurrence of major geological events (e.g. episodes of mountain building or major changes in level of seas).

The earliest era is Precambrian, which began about 4600 million years ago. It has few fossils. Abundant fossils have been reported in the succeeding eras viz. Palaeozoic, Mesozoic and Cenozoic.

Geological Eras

A geological era is a very long division of geological time, lasting tens of millions of years. Its beginning and end are recognized by major changes in layers of rocks and fossils in the earth. As mentioned above, the earliest era is Precambrian It began about 4600 millions of years ago. Its duration has been 4030 millions of years.

Palaeozoic era began about 570 millions of years ago. Mesozoic era began about 225 millions of years ago. Cenozoic is the most recent era. It began about 65 million years ago, after the Mesozoic era, and it is still continuing.

Eras are divided into periods and epochs.

Geological Periods

A geological era is divided further into several subdivisions called periods. So a period is a major subdivision of a geological era. Major periods of Palaeozoic era are Permian, Carboniferous, Devonian, Silurian, Ordovician and Cambrian. Mesozoic era is divided into three periods viz. Crcta ceous, Jurassic and Triassic.

Similarly Cenozoic era is divided into two periods viz. Quarternary and Tertiary. The oldest known period of geological time scale is Cambrian (began about 570 million years ago) while the most recent period is Quaternary (began about 2.5 million years ago).

Periods are divided into epochs.

Geological Epochs

A subdivision of a geological period is called a geological epoch. Recent, Pleistocene, Pliocene, Miocene, Oligocene, Eocene and Paleocene are the epochs of Periods Quaternary and Tertiary of Cenozoic era.

Major Events of Geological Time Scale:

1. Bryophytes evolved on the earth during the Silurian Period of Paleozoic era (i.e. between 395 to 430 million years ago) and are still surviving.
2. Pteridophytes evolved sometime in Silurian, dominated the earth during Carboniferous and are still surviving.
3. Gymnosperms evolved sometime at the end of Triassic Period (i.e. about 225 million years ago) Mesozoic era, dominated the earth sometime during Palaeocene epoch (i.e. about 65 million years ago) and are still surviving.

4. Angiosperms evolved during Jurassic Period of Mesozoic era and are now dominating the earth in the recent epoch of Quaternary Period of Cenozoic era.

Geologic Mapping

Geologic mapping is a highly interpretive, scientific process that can produce a range of map products for many different uses, including assessing ground-water quality and contamination risks; predicting earthquake, volcano, and landslide hazards; characterizing energy and mineral resources and their extraction costs; waste repository siting; land management and land-use planning; and general education.

The public has come to expect near-instantaneous delivery of relevant, understandable information via the Internet, which in turn has begun to affect the methods used in geologic mapping, as well as the nature of the product. Geologists are rapidly incorporating GIS and information technology (IT) techniques into the production and dissemination of geologic maps.

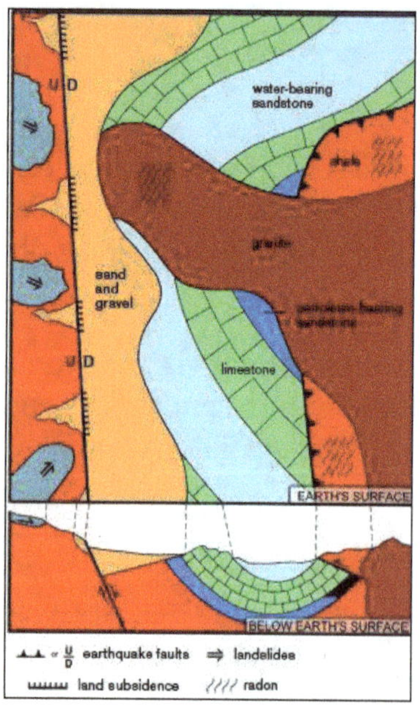

Graphic representation of typical information in a general purpose geologic map. that can be used to identify geologic hazards, locate natural resources, and facilitate land-use planning.

Geologic mapping is a highly interpretive, scientific process that can produce a range of map products for many different uses, including assessing ground-water quality and contamination risks; predicting earthquake, volcano, and landslide hazards; characterizing energy and mineral resources and their extraction costs; waste repository siting; land management and land-use planning; and general education. The value of geologic map information in public and private decision-making (such as for the siting of landfills and highways) has repeatedly been described anecdotally, and has been demonstrated in benefit-cost analyses to reduce uncertainty and, by extension, potential costs.

The geologic mapper strives to understand the composition and structure of geologic materials at the Earth's surface and at depth, and to depict observations and interpretations on maps using symbols and colors. Within the past 10 to 20 years, geographic information system (GIS) technology has begun to change aspects of geologic mapping by providing software tools that permit the geometry and characteristics of rock bodies and other geologic features (such as faults) to be electronically stored, displayed, queried, and analyzed in conjunction with a seemingly infinite variety of other data types.

For example, GIS can be used to spatially compare possible pollutant sources (such as oil wells) with nearby streams and geologic units that serve as ground-water supplies. In addition, GIS can be used to compare the position of a proposed road with the surrounding geology to identify areas of high excavation costs or unstable slopes. These comparisons have always been possible, but GIS greatly facilitates the analysis and, as a result, offers geologists the opportunity to provide information in map form that is easily interpreted and used by the nongeologist.

Geologic Mapping Field Methods

Geologists traditionally record in field notebooks their observations, including sketches, measurements (for example, the angle of tilted strata), and narratives. The validity of these observations remains; however, digital photographs now frequently supplement sketches, and instrumentation enhances measurement accuracy (for example, more precise locations are possible with a global positioning system instrument than with simple reference to position in relation to topographic and cultural features). Increasingly, field narratives are written and organized on a notebook computer or a personal digital assistant (PDA).

The development of field systems for recording and managing information has accelerated in the past 10 years. It is expected that the most useful systems will be widely adopted, thereby helping to standardize the techniques and formats for field observations. Increasingly, geologists not only are recording field observations but also are making field interpretations of the position of features such as geologic contacts and faults (that is, "drawing lines") using rudimentary GIS software on PDAs or notebook computers.

Geologic Map Descriptors

When recording observations, geologists use descriptive terms and rock names that are in common use or unique to an area. These terms are then synthesized and rewritten into formal map unit descriptions that are published with the map. With the advent of GIS and the ease with which digital maps can be obtained and queried, geologists are recognizing the importance of a well defined, standard terminology in order to help users, at a desktop computer or on the Internet, query simultaneously two maps made by different geologists. National and international efforts are now underway to define standard classifications for geologic information such as rock composition and texture. This standard language then can be used in the field or in the office to organize and interpret field observations. It is recognized that because of the many geologic terrains and geologic mapping agencies, and because of long historical usage of certain terms, multiple standard classifications will be necessary to accommodate regional variations in terminology. However, this system should function well, provided each classification is well defined and can be correlated with other classifications to ensure ease of translation from one format to another.

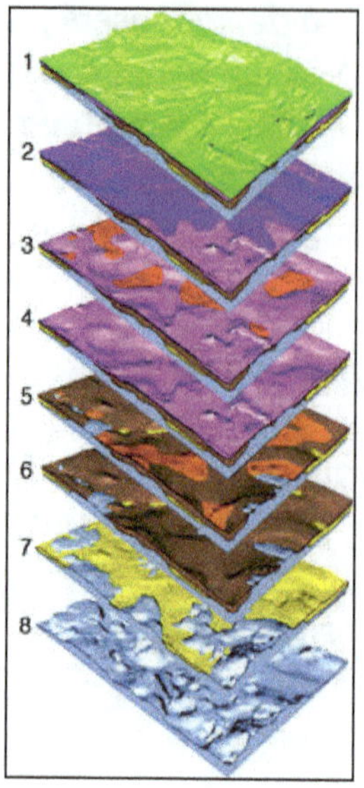

Three-dimensional stack of glacial geologic layers in east-central Illinois. Layer 1 (top) is land surface; layer 8 (bottom) is the underlying bedrock. The light-colored unit in layer 7 is a sand and gravel aquifer filling a bedrock valley, and buried by low-permeability glacial till.

Geologic Map Databases

The main goal in many geological surveys no longer is to create a single geologic map but to create a database from which many types of geologic and engineering geology maps can be derived. This requires a database design or "data model" that is sufficiently robust to manage complex geologic concepts such as three dimensional (spatial) and temporal relations among map units, faults, and other features.

This is especially challenging because the software that manages these databases is not static but continues to evolve, thereby requiring an adaptable database design. Also, to permit the exchange of databases among agencies and users, either a common database design or a common interchange format is required. These design efforts are underway in North America and other parts of the world. As a consequence, geologists now are reevaluating how information is managed in the field, what kind of information is gathered and for what purpose, and the extent to which map information, once it has been created, can be updated to include new observations and interpretations.

Traditionally, once the geologist delivered a manuscript map to the cartographer, the job was finished. Map databases are meant to be maintained, and so a map can now be more accurately considered a progress report that can be updated. A critical aspect of the map database is the ability to manage the information and to preserve its integrity, for example, by migrating it to a new data format or structure or to a new standard terminology.

Cartography and Geologic Maps

GIS and graphic design software have radically changed the techniques by which map information is published. Traditional film (peelcoat) technologies have been supplanted by digitally prepared negatives for offset printing of paper maps and by plot-on-demand approaches, which generate one map at a time upon request by users. Plot-on-demand is useful for producing customized, single-attribute maps (for example, showing only the shear strength of the geologic materials or the susceptibility of a groundwater supply to contamination). Digital cartographic techniques are evolving and becoming capable of producing sophisticated map layouts and products. It is anticipated that digital cartographers will be continually challenged to develop new techniques as software evolves and as geologists and users demand more complex and informative products.

Delivery of Geologic Maps and Databases

In addition to the conventional venues for obtaining maps (such as bookstores and sales offices), GIS and the Internet have made it possible to reach and educate new and potential users of geologic maps. Maps and databases now are available on the Internet in a variety of formats. Some formats (for example, Adobe Portable Document Format, or PDF) are designed for the visual display of a map and do not require specialized software, whereas other formats (for example, ArcInfoTM) are used by professionals with the appropriate software for the detailed analysis of a map database. To provide the public with access to such analyses without requiring them to purchase the software, numerous agencies are experimenting with software that permits users to view maps and to submit queries and view results within a Web browser. This technology is becoming increasingly widespread and should prove ever more useful as the following conditions are met:

1. Increased availability of digital geology information.
2. Standardized database structure, terminology, and interchange format.
3. Increased Internet bandwidth.
4. Advances in the software for serving map data.

New advances in mapping and preparation of map products made possible by advances in GIS and information technology have not altered the basic science of geology but offer new techniques for organizing, maintaining, and analyzing map data, and, potentially, for increasing its use by the public and by scientists.

Digital Geologic Mapping

Screenshot of a structure map generated by geological mapping software for an 8500ft deep gas & Oil reservoir in the Erath field, Vermilion Parish, Erath, Louisiana. The left-to-right gap, near the top of the contour map indicates a Fault line. This fault line is between the blue/green contour lines and the purple/red/yellow contour lines. The thin red circular contour line in the middle of the map indicates the top of the oil reservoir. Because gas floats above oil, the thin red contour line marks the gas/oil contact zone.

Digital geologic mapping is the process by which geological features are observed, analyzed, and recorded in the field and displayed in real-time on a computer or personal digital assistant (PDA). The primary function of this emerging technology is to produce spatially referenced geologic maps that can be utilized and updated while conducting field work.

Traditional Geologic Mapping

Geologic mapping is an interpretive process involving multiple types of information, from analytical data to personal observation, all synthesized and recorded by the geologist. Geologic observations have traditionally been recorded on paper, whether on standardized note cards, in a notebook, or on a map.

Mapping in the Digital Era

In the 21st century, computer technology and software are becoming portable and powerful enough to take on some of the more mundane tasks a geologist must perform in the field, such as precisely locating oneself with a GPS unit, displaying multiple images (maps, satellite images, aerial photography, etc.), plotting strike and dip symbols, and color-coding different physical characteristics of a lithology or contact type (e.g., unconformity) between rock strata. Additionally, computers can now perform some tasks that were difficult to accomplish in the field, for example, handwriting or voice recognition and annotating photographs on the spot.

Digital mapping has positive and negative effects on the mapping process; only an assessment of its impact on a geological mapping project as a whole shows whether it provides a net benefit. With the use of computers in the field, the recording of observations and basic data management changes dramatically. The use of digital mapping also affects when data analysis occurs in the mapping process, but does not greatly affect the process itself.

Advantages

- Data entered by a geologist may have fewer errors than data transcribed by a data entry clerk.
- Data entry by geologists in the field may take less total time than subsequent data entry in the office, potentially reducing the overall time needed to complete a project.
- The spatial extent of real world objects and their attributes can be entered directly into a database with geographic information system (GIS) capability. Features can be automatically color-coded and symbolized based on set criteria.
- Multiple maps and imagery (geophysical maps, satellite images, orthophotos, etc.) can easily be carried and displayed on-screen.
- Geologists may upload each other's data files for the next day's field work as reference.
- Data analysis may start immediately after returning from the field, since the database has already been populated.
- Data can be constrained by dictionaries and dropdown menus to ensure that data are recorded systematically and that mandatory data are not forgotten.
- Labour-saving tools and functionlaity can be provided in the field e.g. structure contours on the fly, and 3D visualisation.
- Systems can be wirelessly connected to other digital field equipment (such as digital cameras and sensor webs).

Disadvantages

- Computers and related items (extra batteries, stylus, cameras, etc.) must be carried in the field.

- Field data entry into the computer may take longer than physically writing on paper, possibly resulting in longer field programs.
- Data entered by multiple geologists may contain more inconsistencies than data entered by one person, making the database more difficult to query.
- Written descriptions convey to the reader detailed information through imagery that may not be communicated by the same data in parsed format.
- Geologists may be inclined to shorten text descriptions because they are difficult to enter (either by handwriting or voice recognition), resulting in loss of data.
- There are no original, hardcopy field maps or notes to archive. Paper is a more stable medium than digital format.

Educational and Scientific Uses

Some universities and secondary educators are integrating digital geologic mapping into class work. For example, The GeoPad project describes the combination of technology, teaching field geology, and geologic mapping in programs such as Bowling Green State University's geology field camp. At Urbino University (Italy) it:Università di Urbino, Field Digital Mapping Techniques are integrated in Earth and Environmental Sciences courses since 2006 . The MapTeach program is designed to provide hands-on digital mapping for middle and high school students. The SPLINT project in the UK is using the BGS field mapping system as part of their teaching curriculum.

Digital mapping technology can be applied to traditional geologic mapping, reconnaissance mapping, and surveying of geologic features. At international digital field data capture (DFDC) meetings, major geological surveys (e.g., British Geological Survey and Geological Survey of Canada) discuss how to harness and develop the technology. Many other geological surveys and private companies are also designing systems to conduct scientific and applied geological mapping of, for example, geothermal springs and mine sites.

Equipment

The initial cost of digital geologic computing and supporting equipment may be significant. In addition, equipment and software must be replaced occasionally due to damage, loss, and obsolescence. Products moving through the market are quickly discontinued as technology and consumer interests evolve. A product that works well for digital mapping may not be available for purchase the following year; however, testing multiple brands and generations of equipment and software is prohibitively expensive.

Common Essential Features

Some features of digital mapping equipment are common to both survey or reconnaissance mapping and "traditional" comprehensive mapping. The capture of less data-intensive reconnaissance mapping or survey data in the field can be accomplished by less robust databases and GIS programs, and hardware with a smaller screen size.

- Devices and software are intuitive to learn and easy to use.

- Rugged, as typically defined by military standards (MIL-STD-810) and ingress protection ratings.
- Waterproof.
- Screen is easy to read in bright sunlight and on gray sky days.
- Removable static memory cards can be used to back up data.
- Memory on board is recoverable.
- Real-time and post-processing differential correction for GPS locations.
- Portable battery with at least 9 hours of life at near constant use.
- Can change batteries in the field.
- Batteries should have no "memory," such as with NiCd.
- Chargeable by unconventional power sources generators, solar, etc.
- Wireless real-time link to GPS or built-in GPS.
- Wireless real-time link from computer to camera and other peripherals
- USB ports.

Features Essential to Capture Traditional Geologic Observations

Hardware and software only recently became available that can satisfy most of the criteria necessary for digitally capturing "traditional" mapping data.

- Screen about 5" × 7"—compact but large enough to see map features. In 2009, some traditional mapping is conducted on PDAs.
- Lightweight—ideally less than 3 lbs.
- Transcription to digital text from handwriting and voice recognition.
- Can store paragraphs of data (text fields).
- Can store complex relational database with drop-down lists.
- Operating system and hardware are compatible with a robust GIS program.
- At least 512 MB memory.

Technology

Software

Since every geologic mapping project covers an area with unique lithologies and complexities, and every geologist has a unique style of mapping, no software is perfect for digital geologic mapping out of the box. The geologist can choose to either modify their mapping style to the available software, or modify the software to their mapping style, which may require extensive programming. As of 2009, available geologic mapping software requires some degree of customization for a given geologic mapping project. Some digital-mapping geologists/programmers have chosen to highly customize or extend ESRI's ArcGIS instead. At digital field data capture meetings such as at the British Geological Survey in 2002 some organisations agreed to share development experiences, and some software systems are now available to download for free.

Radiometric Dating

Radiometric dating, radioactive dating or radioisotope dating is a technique used to date materials such as rocks or carbon, in which trace radioactive impurities were selectively incorporated when they were formed. The method compares the abundance of a naturally occurring radioactive isotope within the material to the abundance of its decay products, which form at a known constant rate of decay. The use of radiometric dating was first published in 1907 by Bertram Boltwoodand is now the principal source of information about the absolute age of rocks and other geological features, including the age of fossilized life forms or the age of the Earth itself, and can also be used to date a wide range of natural and man-made materials.

Together with stratigraphic principles, radiometric dating methods are used in geochronology to establish the geologic time scale. Among the best-known techniques are radiocarbon dating, potassium–argon dating and uranium–lead dating. By allowing the establishment of geological timescales, it provides a significant source of information about the ages of fossilsand the deduced rates of evolutionary change. Radiometric dating is also used to date archaeological materials, including ancient artifacts.

Different methods of radiometric dating vary in the timescale over which they are accurate and the materials to which they can be applied.

Fundamentals

Radioactive Decay

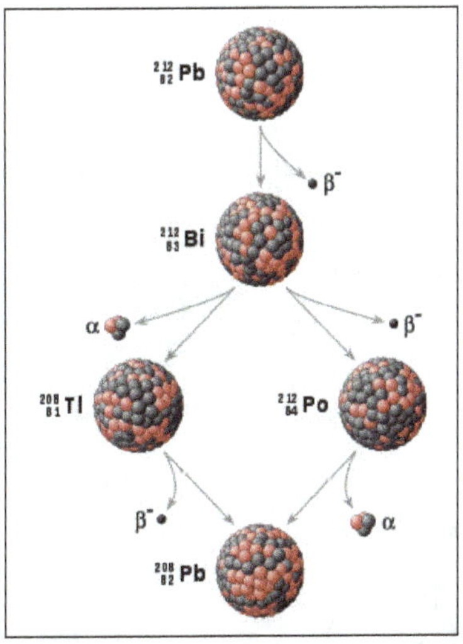

Example of a radioactive decay chain from lead-212 (^{212}Pb) to lead-208 (^{208}Pb) .
Each parent nuclide spontaneously decays into a daughter nuclide (the decay product) via an α decay or a β^- decay.
The final decay product, lead-208 (^{208}Pb), is stable and can no longer undergo spontaneous radioactive decay.

All ordinary matter is made up of combinations of chemical elements, each with its own atomic number, indicating the number of protons in the atomic nucleus. Additionally, elements may exist in different isotopes, with each isotope of an element differing in the number of neutrons in the nucleus. A particular isotope of a particular element is called a nuclide. Some nuclides are inherently unstable. That is, at some point in time, an atom of such a nuclide will undergo radioactive decay and spontaneously transform into a different nuclide. This transformation may be accomplished in a number of different ways, including alpha decay (emission of alpha particles) and beta decay(electron emission, positron emission, or electron capture). Another possibility is spontaneous fission into two or more nuclides.

While the moment in time at which a particular nucleus decays is unpredictable, a collection of atoms of a radioactive nuclide decays exponentially at a rate described by a parameter known as the half-life, usually given in units of years when discussing dating techniques. After one half-life has elapsed, one half of the atoms of the nuclide in question will have decayed into a "daughter" nuclide or decay product. In many cases, the daughter nuclide itself is radioactive, resulting in a decay chain, eventually ending with the formation of a stable (nonradioactive) daughter nuclide; each step in such a chain is characterized by a distinct half-life. In these cases, usually the half-life of interest in radiometric dating is the longest one in the chain, which is the rate-limiting factor in the ultimate transformation of the radioactive nuclide into its stable daughter. Isotopic systems that have been exploited for radiometric dating have half-lives ranging from only about 10 years (e.g., tritium) to over 100 billion years (e.g., samarium-147).

For most radioactive nuclides, the half-life depends solely on nuclear properties and is essentially a constant. It is not affected by external factors such as temperature, pressure, chemical environment, or presence of a magnetic or electric field. The only exceptions are nuclides that decay by the process of electron capture, such as beryllium-7, strontium-85, and zirconium-89, whose decay rate may be affected by local electron density. For all other nuclides, the proportion of the original nuclide to its decay products changes in a predictable way as the original nuclide decays over time. This predictability allows the relative abundances of related nuclides to be used as a clock to measure the time from the incorporation of the original nuclides into a material to the present.

Accuracy of Radiometric Dating

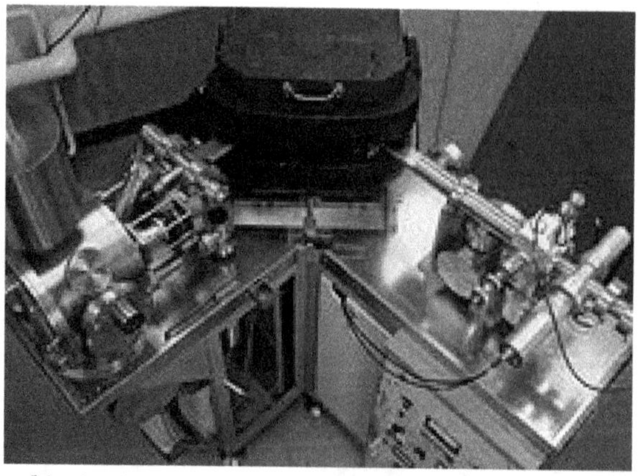

Thermal ionization mass spectrometer used in radiometric dating.

The basic equation of radiometric dating requires that neither the parent nuclide nor the daughter product can enter or leave the material after its formation. The possible confounding effects of contamination of parent and daughter isotopes have to be considered, as do the effects of any loss or gain of such isotopes since the sample was created. It is therefore essential to have as much information as possible about the material being dated and to check for possible signs of alteration. Precision is enhanced if measurements are taken on multiple samples from different locations of the rock body. Alternatively, if several different minerals can be dated from the same sample and are assumed to be formed by the same event and were in equilibrium with the reservoir when they formed, they should form an isochron. This can reduce the problem of contamination. In uranium–lead dating, the concordia diagram is used which also decreases the problem of nuclide loss. Finally, correlation between different isotopic dating methods may be required to confirm the age of a sample. For example, the age of the Amitsoq gneisses from western Greenland was determined to be 3.6 ± 0.05 million years ago (MA) using uranium–lead dating and 3.56 ± 0.10 Ma using lead–lead dating, results that are consistent with each other.

Accurate radiometric dating generally requires that the parent has a long enough half-life that it will be present in significant amounts at the time of measurement (except as described below under "Dating with short-lived extinct radionuclides"), the half-life of the parent is accurately known, and enough of the daughter product is produced to be accurately measured and distinguished from the initial amount of the daughter present in the material. The procedures used to isolate and analyze the parent and daughter nuclides must be precise and accurate. This normally involves isotope-ratio mass spectrometry.

The precision of a dating method depends in part on the half-life of the radioactive isotope involved. For instance, carbon-14 has a half-life of 5,730 years. After an organism has been dead for 60,000 years, so little carbon-14 is left that accurate dating cannot be established. On the other hand, the concentration of carbon-14 falls off so steeply that the age of relatively young remains can be determined precisely to within a few decades.

Closure Temperature

If a material that selectively rejects the daughter nuclide is heated, any daughter nuclides that have been accumulated over time will be lost through diffusion, setting the isotopic "clock" to zero. The temperature at which this happens is known as the closure temperature or blocking temperature and is specific to a particular material and isotopic system. These temperatures are experimentally determined in the lab by artificially resetting sample minerals using a high-temperature furnace. As the mineral cools, the crystal structure begins to form and diffusion of isotopes is less easy. At a certain temperature, the crystal structure has formed sufficiently to prevent diffusion of isotopes. This temperature is what is known as closure temperature and represents the temperature below which the mineral is a closed system to isotopes. Thus an igneous or metamorphic rock or melt, which is slowly cooling, does not begin to exhibit measurable radioactive decay until it cools below the closure temperature. The age that can be calculated by radiometric dating is thus the time at which the rock or mineral cooled to closure temperature. Dating of different minerals and isotope systems (with differing closure temperatures) within the same rock can therefore enable the tracking of the thermal history of the rock in question with time, and thus the history of metamorphic events may become known in detail. This field is known as thermochronology or thermochronometry.

Age Equation

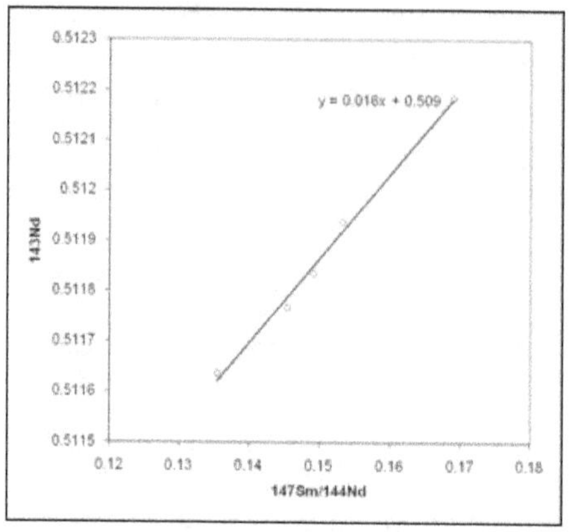

Sm/Nd Isochron plotted of samples from the Great Dyke, Zimbabwe.
The age is calculated from the slope of the isochron (line) and the original composition from the intercept of the isochron with the y-axis.

The mathematical expression that relates radioactive decay to geologic time is:

$$D = D_0 + N(t)\,(e^{\lambda t} - 1)$$

Where,

t is age of the sample,

D is number of atoms of the daughter isotope in the sample,

D_0 is number of atoms of the daughter isotope in the original composition,

N(t) is number of atoms of the parent isotope in the sample at time t (the present), given by $N(t) = N_0 e^{-\lambda t}$, and

λ is the decay constant of the parent isotope, equal to the inverse of the radioactive half-life of the parent isotopetimes the natural logarithm of 2.

The equation is most conveniently expressed in terms of the measured quantity *N(t)* rather than the constant initial value N_0.

The above equation makes use of information on the composition of parent and daughter isotopes at the time the material being tested cooled below its closure temperature. This is well-established for most isotopic systems. However, construction of an isochron does not require information on the original compositions, using merely the present ratios of the parent and daughter isotopes to a standard isotope. Plotting an isochron is used to solve the age equation graphically and calculate the age of the sample and the original composition.

Modern Dating Methods

Radiometric dating has been carried out since 1905 when it was invented by Ernest Rutherford as a method by which one might determine the age of the Earth. In the century since then the techniques have been greatly improved and expanded. Dating can now be performed on samples as small as a nanogram using a mass spectrometer. The mass spectrometer was invented in the 1940s and began to be used in radiometric dating in the 1950s. It operates by generating a beam of ionized atoms from the sample under test. The ions then travel through a magnetic field, which diverts them into different sampling sensors, known as "Faraday cups", depending on their mass and level of ionization. On impact in the cups, the ions set up a very weak current that can be measured to determine the rate of impacts and the relative concentrations of different atoms in the beams.

Uranium–lead Dating Method

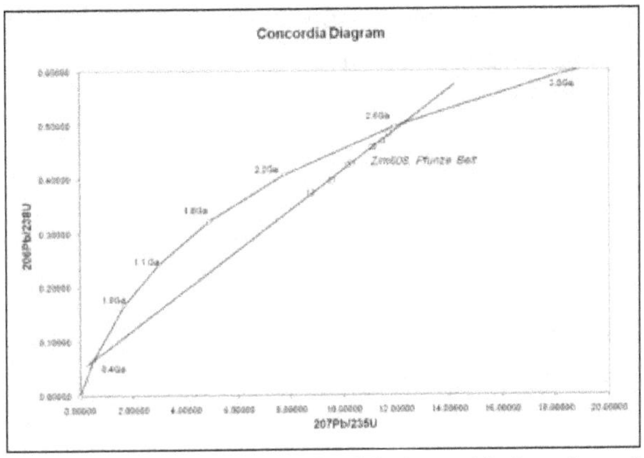

A concordia diagram as used in uranium–lead dating, with data from the Pfunze Belt, Zimbabwe. All the samples show loss of lead isotopes, but the intercept of the errorchron (straight line through the sample points) and the concordia (curve) shows the correct age of the rock.

Uranium–lead radiometric dating involves using uranium-235 or uranium-238 to date a substance's absolute age. This scheme has been refined to the point that the error margin in dates of rocks can be as low as less than two million years in two-and-a-half billion years. An error margin of 2–5% has been achieved on younger Mesozoic rocks.

Uranium–lead dating is often performed on the mineralzircon ($ZrSiO_4$), though it can be used on other materials, such as baddeleyite, as well as monazite. Zircon and baddeleyite incorporate uranium atoms into their crystalline structure as substitutes for zirconium, but strongly reject lead. Zircon has a very high closure temperature, is resistant to mechanical weathering and is very chemically inert. Zircon also forms multiple crystal layers during metamorphic events, which each may record an isotopic age of the event. *In situ* micro-beam analysis can be achieved via laser ICP-MS or SIMS techniques.

One of its great advantages is that any sample provides two clocks, one based on uranium-235's decay to lead-207 with a half-life of about 700 million years, and one based on uranium-238's decay to lead-206 with a half-life of about 4.5 billion years, providing a built-in crosscheck that allows accurate determination of the age of the sample even if some of the lead has been lost. This can be seen in the concordia diagram, where the samples plot along an errorchron (straight line) which intersects the concordia curve at the age of the sample.

Samarium–neodymium Dating Method

This involves the alpha decay of ^{147}Sm to ^{143}Nd with a half-life of 1.06×10^{11} years. Accuracy levels of within twenty million years in ages of two-and-a-half billion years are achievable.

Potassium–argon Dating Method

This involves electron capture or positron decay of potassium-40 to argon-40. Potassium-40 has a half-life of 1.3 billion years, so this method is applicable to the oldest rocks. Radioactive potassium-40 is common in micas, feldspars, and hornblendes, though the closure temperature is fairly low in these materials, about 350 °C (mica) to 500 °C (hornblende).

Rubidium–strontium Dating Method

This is based on the beta decay of rubidium-87 to strontium-87, with a half-life of 50 billion years. This scheme is used to date old igneous and metamorphic rocks, and has also been used to date lunar samples. Closure temperatures are so high that they are not a concern. Rubidium-strontium dating is not as precise as the uranium-lead method, with errors of 30 to 50 million years for a 3-billion-year-old sample.

Uranium–thorium Dating Method

A relatively short-range dating technique is based on the decay of uranium-234 into thorium-230, a substance with a half-life of about 80,000 years. It is accompanied by a sister process, in which uranium-235 decays into protactinium-231, which has a half-life of 32,760 years.

While uranium is water-soluble, thorium and protactinium are not, and so they are selectively precipitated into ocean-floor sediments, from which their ratios are measured. The scheme has a range of several hundred thousand years. A related method is ionium–thorium dating, which measures the ratio of ionium (thorium-230) to thorium-232 in ocean sediment.

Radiocarbon Dating Method

Ale's Stones at Kåseberga, around ten kilometres south east of Ystad, Sweden were dated at 56 CE using the carbon-14 method on organic material found at the site.

Radiocarbon dating is also simply called Carbon-14 dating. Carbon-14 is a radioactive isotope of carbon, with a half-life of 5,730 years(which is very short compared with the above isotopes), and decays into nitrogen. In other radiometric dating methods, the heavy parent isotopes were produced by nucleosynthesis in supernovas, meaning that any parent isotope with a short half-life should be extinct by now. Carbon-14, though, is continuously created through collisions of neutrons generated by cosmic rays with nitrogen in the upper atmosphere and thus remains at a near-constant level on Earth. The carbon-14 ends up as a trace component in atmospheric carbon dioxide (CO_2).

A carbon-based life form acquires carbon during its lifetime. Plants acquire it through photosynthesis, and animals acquire it from consumption of plants and other animals. When an organism dies, it ceases to take in new carbon-14, and the existing isotope decays with a characteristic half-life (5730 years). The proportion of carbon-14 left when the remains of the organism are examined provides an indication of the time elapsed since its death. This makes carbon-14 an ideal dating method to date the age of bones or the remains of an organism. The carbon-14 dating limit lies around 58,000 to 62,000 years.

The rate of creation of carbon-14 appears to be roughly constant, as cross-checks of carbon-14 dating with other dating methods show it gives consistent results. However, local eruptions of volcanoes or other events that give off large amounts of carbon dioxide can reduce local concentrations of carbon-14 and give inaccurate dates. The releases of carbon dioxide into the biosphere as a consequence of industrialization have also depressed the proportion of carbon-14 by a few percent; conversely, the amount of carbon-14 was increased by above-ground nuclear bomb tests that were conducted into the early 1960s. Also, an increase in the solar wind or the Earth's magnetic field above the current value would depress the amount of carbon-14 created in the atmosphere.

Fission Track Dating Method

Apatite crystals are widely used in fission track dating.

This involves inspection of a polished slice of a material to determine the density of "track" markings left in it by the spontaneous fission of uranium-238 impurities. The uranium content of the sample has to be known, but that can be determined by placing a plastic film over the polished slice of the material, and bombarding it with slow neutrons. This causes induced fission of ^{235}U, as opposed to the spontaneous fission of ^{238}U. The fission tracks produced by this process are recorded in the plastic film. The uranium content of the material can then be calculated from the number of tracks and the neutron flux.

This scheme has application over a wide range of geologic dates. For dates up to a few million years micas, tektites (glass fragments from volcanic eruptions), and meteorites are best used. Older materials can be dated using zircon, apatite, titanite, epidote and garnet which have a variable amount of uranium content. Because the fission tracks are healed by temperatures over about 200 °C the technique has limitations as well as benefits. The technique has potential applications for detailing the thermal history of a deposit.

Chlorine-36 Dating Method

Large amounts of otherwise rare ^{36}Cl (half-life ~300ky) were produced by irradiation of seawater during atmospheric detonations of nuclear weapons between 1952 and 1958. The residence time of ^{36}Cl in the atmosphere is about 1 week. Thus, as an event marker of 1950s water in soil and ground water, ^{36}Cl is also useful for dating waters less than 50 years before the present. ^{36}Cl has seen use in other areas of the geological sciences, including dating ice and sediments.

Luminescence Dating Methods

Luminescence dating methods are not radiometric dating methods in that they do not rely on abundances of isotopes to calculate age. Instead, they are a consequence of background radiation on certain minerals. Over time, ionizing radiation is absorbed by mineral grains in sediments and archaeological materials such as quartz and potassium feldspar. The radiation causes charge to remain within the grains in structurally unstable "electron traps". Exposure to sunlight or heat releases these charges, effectively "bleaching" the sample and resetting the clock to zero. The trapped charge accumulates over time at a rate determined by the amount of background radiation at the location where the sample was buried. Stimulating these mineral grains using either light (optically stimulated luminescence or infrared stimulated luminescence dating) or heat (thermoluminescence dating) causes a luminescence signal to be emitted as the stored unstable electron energy is released, the intensity of which varies depending on the amount of radiation absorbed during burial and specific properties of the mineral.

These methods can be used to date the age of a sediment layer, as layers deposited on top would prevent the grains from being "bleached" and reset by sunlight. Pottery shards can be dated to the last time they experienced significant heat, generally when they were fired in a kiln.

Other Methods

Other methods include:

- Argon–argon (Ar–Ar),
- Iodine–xenon (I–Xe),
- Lanthanum–barium (La–Ba),
- Lead–lead (Pb–Pb),
- Lutetium–hafnium (Lu–Hf),
- Potassium–calcium (K–Ca),
- Rhenium–osmium (Re–Os)
- Uranium–thorium helium (U–Th-He),
- Uranium–uranium (U–U),
- Krypton–krypton (Kr–Kr),
- Beryllium (^{10}Be–^{9}Be).

Dating with Decay Products of Short-lived Extinct Radionuclides

Absolute radiometric dating requires a measurable fraction of parent nucleus to remain in the sample rock. For rocks dating back to the beginning of the solar system, this requires extremely long-lived parent isotopes, making measurement of such rocks' exact ages imprecise. To be able to distinguish the relative ages of rocks from such old material, and to get a better time resolution than that available from long-lived isotopes, short-lived isotopes that are no longer present in the rock can be used.

At the beginning of the solar system, there were several relatively short-lived radionuclides like ^{26}Al, ^{60}Fe, ^{53}Mn, and ^{129}I present within the solar nebula. These radionuclides—possibly produced by the explosion of a supernova—are extinct today, but their decay products can be detected in very old material, such as that which constitutes meteorites. By measuring the decay products of extinct radionuclides with a mass spectrometer and using isochronplots, it is possible to determine relative ages of different events in the early history of the solar system. Dating methods based on extinct radionuclides can also be calibrated with the U-Pb method to give absolute ages. Thus both the approximate age and a high time resolution can be obtained. Generally a shorter half-life leads to a higher time resolution at the expense of timescale.

^{129}I – ^{129}Xe Chronometer

^{129}I beta-decays to ^{129}Xe with a half-life of 16 million years. The iodine-xenon chronometer is an isochron technique. Samples are exposed to neutrons in a nuclear reactor. This converts the only stable isotope of iodine (^{127}I) into ^{128}Xe via neutron capture followed by beta decay (of ^{128}I). After irradiation, samples are heated in a series of steps and the xenon isotopic signature of the gas evolved in each step is analysed. When a consistent ^{129}Xe/^{128}Xe ratio is observed across several consecutive temperature steps, it can be interpreted as corresponding to a time at which the sample stopped losing xenon.

Samples of a meteorite called Shallowater are usually included in the irradiation to monitor the conversion efficiency from ^{127}I to ^{128}Xe. The difference between the measured ^{129}Xe/^{128}Xe ratios of the sample and Shallowater then corresponds to the different ratios of ^{129}I/^{127}I when they each stopped losing xenon. This in turn corresponds to a difference in age of closure in the early solar system.

^{26}Al – ^{26}Mg Chronometer

Another example of short-lived extinct radionuclide dating is the ^{26}Al – ^{26}Mg chronometer, which can be used to estimate the relative ages of chondrules. ^{26}Al decays to ^{26}Mg with a half-life of 720 000 years. The dating is simply a question of finding the deviation from the natural abundance of ^{26}Mg (the product of ^{26}Al decay) in comparison with the ratio of the stable isotopes ^{27}Al/^{24}Mg.

The excess of ^{26}Mg (often designated ^{26}Mg*) is found by comparing the ^{26}Mg/^{27}Mg ratio to that of other Solar System materials.

The ^{26}Al – ^{26}Mg chronometer gives an estimate of the time period for formation of primitive meteorites of only a few million years (1.4 million years for Chondrule formation).

Geophysical Survey

Geophysical survey is the systematic collection of geophysical data for spatial studies. Detection and analysis of the geophysical signals forms the core of Geophysical signal processing. The magnetic and gravitational fields emanating from the Earth's interior hold essential information concerning seismic activities and the internal structure. Hence, detection and analysis of the electric and Magnetic fields is very crucial. As the Electromagnetic and gravitational waves are multi-dimensional signals, all the 1-D transformation techniques can be extended for the analysis of these signals as well. Hence this topic also discusses multi-dimensional signal processing techniques.

Geophysical surveys may use a great variety of sensing instruments, and data may be collected from above or below the Earth's surface or from aerial, orbital, or marine platforms. Geophysical surveys have many applications in geology, archaeology, mineral and energy exploration, oceanography, and engineering. Geophysical surveys are used in industry as well as for academic research.

The sensing instruments such as gravimeter, gravitational wave sensor and magnetometers detect fluctuations in the gravitational and magnetic field. The data collected from a geophysical survey is analysed to draw meaningful conclusions out of that. Analysing the spectral density and the time-frequency localisation of any signal is important in applications such as oil exploration and seismography.

Types of Geophysical Survey

There are many methods and types of instruments used in geophysical surveys. Technologies used for geophysical surveys include:

1. Seismic methods, such as reflection seismology, seismic refraction, and seismic tomography.
2. Seismoelectrical method.
3. Geodesy and gravity techniques, including gravimetry and gravity gradiometry.
4. Magnetic techniques, including aeromagnetic surveys and magnetometers.
5. Electrical techniques, including electrical resistivity tomography, induced polarization, spontaneous potential and marine control source electromagnetic (mCSEM) or EM seabed logging.
6. Electromagnetic methods, such as magnetotellurics, ground penetrating radar and transient/time-domain electromagnetics, surface nuclear magnetic resonance (also known as magnetic resonance sounding).
7. Borehole geophysics, also called well logging.
8. Remote sensing techniques, including hyperspectral.

Geophysical Signal Detection

This section deals with the principles behind measurement of geophysical waves. The magnetic and gravitational fields are important components of geo-physical signals. Hence, a brief discussion is important. This section also deals with the equipment used in detecting the magnetic and gravitational waves.

Measurement of Earth's Magnetic Fields

Magnetometers are used to measure the magnetic fields, magnetic anomalies in the earth. The sensitivity of magnetometers depends upon the requirement. For example, the variations in the geomagnetic fields can be to the order of several aT where $1aT = 10^{-18}T$. In such cases, specialized magnetometers such as the superconducting quantum interference device(SQUID) are used.

SQUID

Jim Zimmerman co-developed the rf SQUID (Superconducting QUantum Interference Device) during his tenure at Ford research lab. However, events leading to the invention of SQUID were in fact, serendipitous. John Lambe, during his experiments on Nuclear Magnetic Resonance noticed that the electrical properties of Indium varied due to a change in the magnetic field of the order of few nT. But, Lambe was not able to fully recognise the utility of SQUID.

SQUIDs have the capability to detect magnetic fields of extremely low magnitude. This is due to the virtue of the Josephson junction. Jim Zimmerman pioneered the development of SQUID by proposing a new approach to making the Josephson junctions. He made use of niobium wires and Niobium ribbons to form two Josephson junctions connected in parallel. The ribbons act as the interruptions to the superconducting current flowing through the wires. The junctions are very sensitive to the magnetic fields and hence are very useful in measuring fields of the order of $10^{\wedge -18}T$.

Seismic Wave Measurement using Gravitational Wave Sensor

Gravitational wave sensors can detect even a minute change in the gravitational fields due to the influence of heavier bodies. Large seismic waves can interfere with the gravitational waves and may cause shifts in the atoms. Hence, the magnitude of seismic waves can be detected by a relative shift in the gravitational waves.

Measurement of Seismic Waves using Atom Interferometer

The motion of any mass is affected by the gravitational field. The motion of planets is affected by the Sun's enormous gravitational field. Likewise, a heavier object will influence the motion of other objects of smaller mass in its vicinity. However, this change in the motion is very small compared to the motion of heavenly bodies. Hence, special instruments are required to measure such a minute change.

Atom interferometers work on the principle of diffraction. The diffraction gratings are nano fabricated materials with a separation of a quarter wavelength of light. When a beam of atoms pass through a diffraction grating, due the inherent wave nature of atoms, they split and form interference fringes on the screen. An atom interferometer is very sensitive to the changes in the positions of atoms.As heavier objects shifts the position of the atoms nearby, displacement of the atoms can be measured by detecting a shift in the interference fringes.

Existing Approaches in Geophysical Signal Recognition

This topic addresses the methods and mathematical techniques behind signal recognition and signal analysis. It considers the time domain and frequency domain analysis of signals. This topic also discusses various transforms and their usefulness in the analysis of multi-dimensional waves.

3D Sampling

The first step in any signal processing approach is analog to digital conversion. The geophysical signals in the analog domain has to be converted to digital domain for further processing.

Analog to Digital Conversion

As the name suggests, the gravitational and electromagnetic waves in the analog domain are detected, sampled and stored for further analysis. The signals can be sampled in both time and frequency domains. The signal component is measured at both intervals of time and space. Ex, time-domain sampling refers to measuring a signal component at several instances of time. Similarly, spatial-sampling refers to measuring the signal at different locations in space.

Traditional sampling of 1D time varying signals is performed by measuring the amplitude of the signal under consideration in discrete intervals of time. Similarly sampling of space-time signals (signals which are functions of 4 variables – 3D space and time), is performed by measuring the amplitude of the signals at different time instances and different locations in the space. For example, the earth's gravitational data is measured with the help of gravitational wave sensor or gradiometer by placing it in different instances at different instances of time.

Spectrum Analysis

Multi-dimensional Fourier Transform

The Fourier expansion of a time domain signal is the representation of the signal as a sum of its frequency components, specifically sum of sines and cosines. Joseph Fourier came up with the Fourier representation to estimate the heat distribution of a body. The same approach can be followed to analyse the multi-dimensional signals such as gravitational waves and electromagnetic waves.

The 4D Fourier representation of such signals is given by:

$$S(K,\omega) = \iint s(x,t)e^{-j(\omega t - k'x)}\,dx\,dt$$

- ω represents temporal frequency and k represents spatial frequency.
- $s(x,t)$ is a 4-dimensional space-time signal which can be imagined as travelling plane waves. For such plane waves, the plane of propagation is perpendicular to the direction of propagation of the considered wave.

Wavelet Transform

The motivation for development of the Wavelet transform was the Short-time Fourier transform.

The signal to be analysed, say *f*(*t*) is multiplied with a window function $w(t)$ at a particular time instant. Analysing the Fourier coefficients of this signal gives us information about the frequency components of the signal at a particular time instant.

The STFT is mathematically written as:

$$\{x(t)\}(\tau,\omega) \equiv X(\tau,\omega) = \int_{-\infty}^{\infty} x(t)w(t-\tau)e^{-j\omega t}\,dt$$

The Wavelet transform is defined as:

$$X(a,b) = \frac{1}{\sqrt{a}}\int \Psi\left(\frac{t-b}{a}\right)x(t)dt$$

A variety of window functions can be used for analysis. Wavelet functions are used for both time and frequency localisation. For example,one of the windows used in calculating the Fourier coefficients is the Gaussian window which is optimally concentrated in time and frequency. This optimal nature can be explained by considering the time scaling and time shifting parameters a and b respectively. By choosing the appropriate values of a and b, we can determine the frequencies and the time associated with that signal. By representing any signal as the linear combination of the wavelet functions, we can localize the signals in both time and frequency domain. Hence wavelet transforms are important in geophysical applications where spatial and temporal frequency localisation is important.

Time Frequency Localization using Wavelets

Geophysical signals are continuously varying functions of space and time. The wavelet transform techniques offer a way to decompose the signals as a linear combination of shifted and scaled version of basis functions. The amount of "shift" and "scale" can be modified to localize the signal in time and frequency.

Beamforming

Simply put, space-time signal filtering problem can be thought as localizing the speed and direction of a particular signal. The design of filters for space-time signals follows a similar approach as that of 1D signals. The filters for 1-D signals are designed in such a way that if the requirement of the filter is to extract frequency components in a particular non-zero range of frequencies, a bandpass filter with appropriate passband and stop band frequencies in determined. Similarly, in the case of multi-dimensional systems, the wavenumber-frequency response of filters is designed in such a way that it is unity in the designed region of (k, ω) a.k.a. wavenumber – frequency and zero elsewhere.

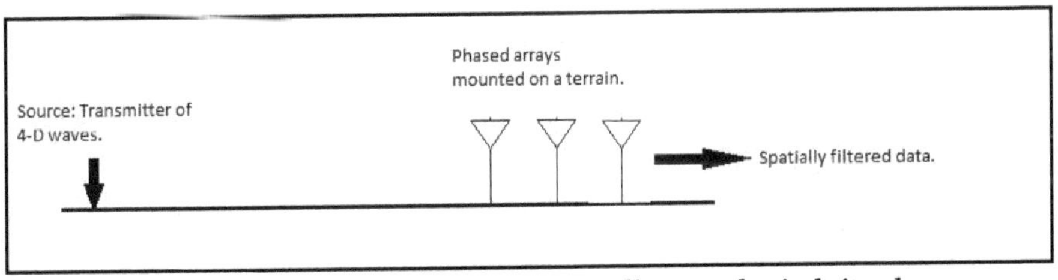

Spatial distribution of phased arrays to filter geophysical signals.

This approach is applied for filtering space-time signals. It is designed to isolate signals travelling in a particular direction. One of the simplest filters is weighted delay and sum beamformer. The output is the average of the linear combination of delayed signals. In other words, the beamformer output is formed by averaging weighted and delayed versions of receiver signals. The delay is chosen such that the passband of beamformer is directed to a specific direction in the space.

Classical Estimation Theory

This deals with the estimation of the power spectral density of the multi-dimensional signals. The spectral density function can be defined as a multidimensional Fourier transform of the autocorrelation function of the random signal.

$$P(K_x, w) = \int_{-\infty}^{\infty}\int_{-\infty}^{\infty} \varphi_{ss}(x,t)\, e^{-j(wt-k'x)}\, dx\, dt\sqrt{2}$$

$$\varphi_{ss}(x,t) = s\left[(\xi,\tau)s^*(\xi - x, \tau - t)\right]$$

The spectral estimates can be obtained by finding the square of the magnitude of the Fourier transform also called as Periodogram. The spectral estimates obtained from the periodogram have a large variance in amplitude for consecutive periodogram samples or in wavenumber. This problem is resolved using techniques that constitute the classical estimation theory. They are as follows:

1. Bartlett suggested a method that averages the spectral estimates to calculate the power spectrum. Average of spectral estimates over a time interval gives a better estimate.

$$P_B(w) = \frac{1}{\det N}\sum_{l}\left|\sum_{n} x(n+Ml)\, e^{-j(w'n)}\right|^2 \text{ Bartlett's case}$$

2. Welch's method suggested to divide the measurements using data window functions, calculate a periodogram, average them to get a spectral estimate and calculate the power spectrum using Fast Fourier Transform (FFT).This increased the computational speed.

$$P_W(w) = \frac{1}{\det N}\sum_{l}\left|\sum_{n} g(n)\, x(n+Ml)\, e^{-j(w'n)}\right|^2 \text{ Welch's case}$$

4. The periodogram under consideration can be modified by multiplying it with a window function. Smoothing window will help us smoothen the estimate. Wider the main lobe of the smoothing spectrum, smoother it becomes at the cost of frequency resolution.

$$P_M(w) = \frac{1}{\det N}\left|\sum_{n} g(n)\, x(n)\, e^{-j(w'n)}\right|^2 \text{ Modified periodogram.}$$

Applications

Estimating Positions of Underground Objects

The method being discussed here assumes that the mass distribution of the underground objects of interest is already known and hence the problem of estimating their location boils down to parametric localisation.Say underground objects with center of masses (CM_1, CM_2...CM_n) are located under the surface and at positions p_1, p_2...p_n The gravity gradient(components of the gravity field) is measured using a spinning wheel with accelerometers also called as the gravity gradiometer. The instrument is positioned in different orientations to measure the respective component of the gravitational field. The values of gravitational gradient tensors are calculated and analyzed. The analysis includes observing the contribution of each object under consideration. A maximum likelihood procedure is followed and Cramér–Rao bound (CRB) is computed to assess the quality of location estimate.

Array Processing for Seismographic Applications

Various sensors located on the surface of earth spaced equidistantly receive the seismic waves. The seismic waves travel through the various layers of earth and undergo changes in their properties - amplitude change, time of arrival, phase shift. By analyzing these properties of the signals, we can model the activities inside the earth.

Visualization of 3D data

The method of volume rendering is an important tool to analyse the scalar fields. Volume rendering simplifies representation of 3D space. Every point in a 3D space is called a voxel. Data inside the 3-d dataset is projected to the 2-d space(display screen) using various techniques. Different data encoding schemes exist for various applications such as MRI, Seismic applications.

Applied Geology 7

- Economic Geology
- Engineering Geology
- Environmental Geology
- Planetary Geology
- Geomicrobiology
- Forensic Geology
- Geologic Modeling

The domain of geology which makes use of the principles and methods of pure geology to fulfill various material needs of humans is referred to applied geology. It encompasses diverse fields such as economic geology, engineering geology, environmental geology, forensic geology and geologic modeling. This chapter closely examines these subdomains of applied geology to provide an extensive understanding of the subject.

Economic Geology

Economic geology is a branch of geology that deals with economically valuable geological materials. In broader terms, economic geology is concerned with the distribution of mineral deposits, the economic considerations involved in their recovery, and assessment of the reserves available.

Economic Geology deals with materials such as precious and base metals, non-metallic minerals, fossil fuels, and other materials of commercial value, such as salt, gypsum, and building stone. It makes use of the principles and methods of various other fields, especially geophysics, structural geology, and stratigraphy.

Economic geology is not only practised by geologists but it is also of some interest to engineers, investment bankers, environmental scientists and conservationists because of the huge impact extractive industries have on the socioeconomic and environmental scene.

The concept of economic geology is a relatively new one, even though humans have been extracting metals and minerals of value from the ground since prehistoric times. For all their ability to appreciate the worth of such resources, however, pre-modern people possessed little in the way of scientific theories regarding either their formation or the means of extracting them.

The Greeks, for instance, believed that veins of metallic materials in the earth indicated that those materials were living things putting down roots after the manner of trees. Astrologers of medieval times maintained that each of the "seven planets" (Sun, Moon, and the five planets, besides Earth, known at the time) ruled one of the seven known metals—gold, copper, silver, lead, tin, iron, and mercury—which supposedly had been created under the influence of their respective "planets."

The first thinker who attempted to go beyond such unscientific (if imaginative) ideas was a German physician writing under the Latinised name Georgius Agricola. As a result of treating miners for various conditions, Agricola, whose real name was Georg Bauer, became fascinated with minerals.

Regarded as the father of both mineralogy and economic geology, Agricola introduced several ideas that provided a scientific foundation for the study of Earth and its products. In De Ortu et causis Subterraneorum, the critiqued all preceding ideas regarding the formation of ores, including the Greek and astrological notions mentioned earlier as well as the alchemical belief that all metals are composed of mercury and sulphur.

Instead, he maintained that subterranean fluids carry dissolved minerals, which, when cooled, leave deposits in the cracks of rocks and thus give rise to mineral veins. Agricola's ideas later helped form the basis for modern theories regarding the formation of ore deposits.

In De Natura Fossilium, Agricola also introduced a method for the classification of "fossils", as minerals were then known. Agricola's system, which categorises minerals according to such properties as colour, texture, weight, and transparency, is the basis for the system of mineral classification in use today.

Of all his works, however, the most important was De re Metallica, which would remain the leading textbook for miners and mineralogists during the two centuries that followed. In this monumental work, he introduced many new ideas, including the concept that rocks contain ores that are older than the rocks themselves. He also explored in detail the mining practices in use in his time, itself an extraordinary feat in that miners of the sixteenth century tended to guard their trade secrets closely.

Rocks and Minerals

Our earth's crust is made up of rocks which, in turn, are aggregates of minerals. To be designated as a mineral species, a substance must be found in nature and be of inorganic origin. It must possess a definite chemical characteristic and a distinctive atomic formation.

Rocks

A rock is an aggregate of minerals or organic material, which can appear in consolidated or unconsolidated form. Rocks are of three different types: igneous, formed by crystallisation of molten minerals, as in a volcano; sedimentary, usually formed by deposition, compaction, or cementation of weathered rock; and metamorphic, formed by alteration of pre-existing rock. Rocks made from organic material are typically sedimentary, an example being coal.

Rocks have possessed economic importance from a time long before 'economics' as we know it existed—a time when there was nothing to buy and nothing to sell. That time, of course, would be the Stone Age, which dates back practically to the beginnings of the human species and overlapped with the beginnings of civilization some 5,500 years ago. In the hundreds of thousands of years when stone constituted the most advanced tool making material, humans developed an array of stone devices for making fire, sharpening knives, killing animals (and other humans), cutting food or animal skins, and so on.

The Stone Age, both in the popular imagination and (with some qualifications) in actual archaeological fact, was a time when people lived in caves. Since that time, of course, humans have generally departed from the caves, though exceptions exist, as the United States military found in 2001 when attempting to hunt for terrorists in the caves of Afghanistan.

In any case, the human attachment to stone dwellings has taken other forms, beginning with the pyramids and continuing through today's masonry homes. Nor is rock simply a structural material for building, as the use of gypsum wallboard, slate countertops, marble finishes, and graveled walkways attests. And, of course, construction is only one of many applications to which rocks and minerals are directed, as we shall see.

Metals

Of all known chemical elements, 87, or about 80 per cent, are metals. The latter group is identified as being lustrous or shiny in appearance and malleable or ductile, meaning that they can be moulded into different shapes without breaking them. Despite their ductility, metals are extremely durable, have high melting and boiling points, and are excellent conductors of heat and electricity. Some register high on the Mohs hardness scale.

Minerals

Whereas there are only 87 varieties of metal, there are some 3,700 types of mineral. There is considerable overlap between metals and minerals, but that overlap is far from complete: many minerals include nonmetallic elements, such as oxygen and silicon. A mineral is a substance that appears in nature and therefore cannot be created artificially, is inorganic in origin, has a definite chemical composition, and possesses a crystalline internal structure.

The term organic does not refer simply to substances with a biological origin; rather, it describes any compound that contains carbon, with the exception of carbonates (which are a type of mineral) and oxides, such as carbon dioxide or carbon monoxide.

The fact that a mineral must be of non-varying composition limits minerals almost exclusively to elements and compounds—that is, either to substances that cannot be broken down chemically to yield simpler substances or to substances formed by the chemical bonding of elements. Only in a few highly specific circumstances are naturally occurring alloys, or mixtures of metals, considered minerals.

Minerals are classified into eight basic groups as per their chemical composition. These are given below:

1. Native elements,
2. Sulphides,
3. Oxides and hydroxides,
4. Halides,
5. Carbonates, nitrates, borates, iodates,
6. Sulphates, chromates, molybdates, tungstates,
7. Phosphates, arsenates, vanadates,
8. Silicates.

The first group, native elements, includes metallic elements that appear in pure form somewhere on Earth; certain metallic alloys, alluded to earlier; as well as native nonmetals, semi-metals, and minerals occur with metallic and nonmetallic elements. The native elements, along with the six classes that follow them in this list, are collectively known as nonsilicates, a term that emphasises the importance of the eighth group.

The vast majority of minerals, including the most abundant ones, belong to the silicates class, which is built around the element silicon. Just as carbon can form long strings of atoms, particularly in combination with hydrogen (as we discuss in the context of fossil fuels later in this essay), silicon also forms long strings, though its "partner of choice" is typically oxygen rather than hydrogen. Together with oxygen, silicon—known as a metalloid because it exhibits characteristics of both metals and nonmetals—forms the basis for an astonishing array of products, both natural and man-made.

Minerals may be Classified according to their use in industry as follows:

1. Metallic minerals: Ferrous group. They include minerals like iron, chromite, manganese and nickel.
2. Metallic minerals: Non-ferrous group. These feature copper, lead, zinc, tungsten, aluminium, vanadium and others.
3. Non-metallic minerals. They are mica, steatite, asbestos and others.
4. Refractory minerals. They are used as heat resistant's in furnaces and moulds. They include chromite, magnesite, kyanite, fireclays, sillimanite and graphite.
5. Fertilizer minerals such as gypsum, rock phosphate and pyrite.
6. Mineral fuels such as coal, petroleum, natural gas and nuclear minerals.

The economic development of a country is influenced by the availability of minerals. Minerals form the base for several large-scale industries. Agriculture, too, is influenced by the availability of minerals in the form of fertilizers.

Hydrocarbons

As noted earlier, the focus of economic geology is on both rocks and minerals, on the one hand, and fossil fuels, on the other. Fossil fuels may be defined as fuels (specifically, coal, oil, and gas) derived from deposits of organic material that have experienced decomposition and chemical alteration under conditions of high pressure.

Given this derivation from organic material, by definition, all fossil fuels are carbon- based, and, specifically, they are built around hydrocarbons—chemical compounds whose molecules are made up of nothing but carbon and hydrogen atoms.

Theoretically, there is no limit to the number of possible hydrocarbons. Carbon forms itself into apparently limitless molecular shapes, and hydrogen is a particularly versatile chemical partner. Hydrocarbons may form straight chains, branched chains, or rings, and the result is a variety of compounds distinguished not by the elements in their makeup or even (in some cases) by the numbers of different atoms in each molecule, but rather by the structure of a given molecule.

Real-life Applications of Economic Geology

Fossil Fuels

The organic material that has decomposed to create the hydrocarbons in fossil fuels comes primarily from dinosaurs and prehistoric plants, though it just as easily could have come from any other organisms that died in large numbers a long, long time ago. To form petroleum, there must be very large quantities of Organic material deposited along with sediments and buried under more sediment. The accumulated sediments and organic material are called source rock.

What happens after accumulation of this material is critical and depends a great deal on the nature of the source rock. It is important that the organic material—for example, the vast numbers of dinosaurs that died in a mass extinction about 65 million years ago—not be allowed simply to rot, as would happen in an aerobic, or oxygen-containing, environment. Instead, the organic material undergoes transformation into hydrocarbons as a result of anaerobic chemical activity, or activity that takes place in the absence of oxygen.

Good source rocks for this transformation are shale or limestone, provided the particular rocks are composed of between 1 per cent and 5 per cent organic carbon. The source rocks should be deep enough that the pressure heats the organic material, yet not so deep that the pressure and temperature cause the rocks to undergo metamorphism or transform them into graphite or other non-hydrocarbon versions of carbon. Temperatures of up to 302 °F (150 °C) are considered optimal for petroleum generation.

Once generated, petroleum gradually moves from the source rock to a reservoir rock or a rock that stores petroleum in its pores. A good reservoir rock is one in which the pore space constitutes more than 30 per cent of the rock volume. Yet the rock must be sealed by another rock that is much less porous; indeed, for a seal or cap rock, as it is called, a virtually impermeable rock is preferred. Thus, the best kind of seal-forming rock is one made of very small, closely fitting pieces of sediment, for instance shale. Such a rock is capable of holding petroleum in place for millions of years until it is ready to be discovered and used.

People have known about petroleum from prehistory, simply because there were places on Earth where it literally seeped from the ground. The modern era of petroleum drilling, however, began in 1853, when an American lawyer named George Bissell recognised its potential for use as a lamp fuel. He hired 'Colonel' Edwin Drake to oversee the drilling of an oil well at Titusville, Pennsylvania, and in 1859 Drake struck oil. The legend of Wack gold', of fortunes to be made by drilling holes in the ground, was born.

In the wake of the development and widespread application of the internal-combustion engine during the latter part of the nineteenth and the early part of the twentieth century's, interest in oil became much more intense, and wells sprouted up around the world. Sumatra, Indonesia yielded- oil from its first wells in 1885, and in 1901, successful drilling began in Texas—the source of many a Texas-sized fortune. An early form of the company known today as British Petroleum (BP) discovered the first Middle Eastern oil in Persia (now Iran) in 1908. Over the next 50 years, the economic importance and prospects of that region changed considerably.

With the vast expansion in automobile ownership that began following World War I and reached even greater heights after World War II, the value and importance of petroleum soared. The oil industry boomed, and, as a result, many geologists found employment in a sector that offered far more in the way of financial benefits than university or government positions ever could. Today geologists assist their employers in locating oil reserves, not an easy task because so many variables must line up to produce a viable oil source. Given the cost of drilling a new oil well, which may run to $30 million or more, it is clearly important to exercise good judgments in assessing the possibilities of finding oil.

The oil industry has been fraught with environmental concerns over the impact of drilling (much of which takes place offshore, on rigs placed in the ocean); possible bio-hazards associated with spills, such as the one involving the Exxon Valdez in 1989; and the effect on the atmosphere of carbon monoxide and other greenhouse gases produced by petroleum-burning internal- combustion engines. There is even more wide- ranging concern over United States' dependence on oil sources in foreign countries (some of which are openly hostile to the United States) as well as the possible dwindling of resources.

At the present rate of consumption, oil reserves are estimated to be exhausted by about the year 2040, but this takes into account only reserves that are considered viable today. As exploration continues, more resources may be tapped. In the long run, however, it will be necessary to develop new means of fuelling the industrialised world, because petroleum is a non-renewable resource: there is only so much of it underground, and when it is gone, it will not be replaced for millions of years.

Petrochemicals

Petroleum itself is a raw material from which numerous products, collectively known as petrochemicals or petroleum derivatives, are obtained. Through a process termed fractional distillation, the petrochemicals of the lowest molecular mass boil off first, and those having higher mass separate at higher temperatures.

Silicon, Silicates and other Compounds

Just as carbon is at the centre of a vast world of hydrocarbons, so silicon is equally important to inorganic substances ranging from sand or silica (SiO_2) to silicone (a highly versatile set of silicon-based products), to the rocks known as silicates.

Silicates are the basis for several well-known mineral types, including garnet, topaz, zircon, kaolinite, talc, mica, and the two most abundant minerals on Earth, feldspar and quartz. (Note that most of the terms used here refer to a group of minerals, not to a single mineral.) Made of compounds formed around silicon and oxygen and comprising various metals, such as aluminum, iron, sodium, and potassium, the silicates account for 30 per cent of all minerals. As such, they appear in everything from gemstones to building materials; yet they are far from the only notable products centered on silicon.

Silicone and other Compounds

Silicone is not a mineral; rather, it is a synthetic product often used as a substitute for organic oils, greases, and rubber. Instead of attaching to oxygen atoms, as in a silicate, silicon atoms in silicone attach to organic groups, that is, molecules containing carbon. Silicone oils frequently are used in place of organic petroleum as a lubricant because they can withstand greater variations in temperature.

And because the body tolerates the introduction of silicone implants better than it does organic ones, silicones are used in surgical implants as well. Silicone rubbers appear in everything from bouncing balls to space vehicles, and silicones are also present in electrical insulators, rust preventives, fabric softeners, hair sprays, hand creams, furniture and automobile polishes, paints, adhesives, and even chewing gum.

Even this list does not exhaust the many applications of silicon, which (together with oxygen) accounts for the vast majority of the mass in Earth's crust. Owing to its semi-metallic qualities, silicon is used as a semiconductor of electricity.

Computer chips are tiny slices of ultra-pure silicon, etched with as many as half a million microscopic and intricately connected electronic circuits. These chips manipulate voltages using binary codes, for which 1 means "voltage on" and 0 means "voltage off." By means of these pulses, silicon chips perform multitudes of calculations in seconds—calculations that would take humans hours or months or even years.

A porous form of silica known as silica gel absorbs water vapour from the air and is often packed alongside moisture-sensitive products, such as electronics components, to keep them dry. Silicon carbine, an extremely hard crystalline material manufactured by fusing sand with coke (almost pure carbon) at high temperatures, has applications as an abrasive.

Ores

An ore is a rock or mineral that possesses economic value. But a more targeted definition would include the adjective metalliferous, since economically valuable minerals that contain no metals usually are treated as a separate category—industrial minerals. Indeed, it can be said that the interests of economic geology are divided into three areas: ores, industrial minerals, and fuels, which we have discussed already.

The very word 'ore' seems to call to mind one of the oldest-known metals in the world and probably the first material worked by prehistoric metallurgists: gold. Even the Spanish word for gold, oro, suggests a connection. When conquistadors from Spain arrived in the New World after about 1500, oro was their obsession, and it was said that the Spanish invaders of Mexico found every bit of gold or silver ore located on the surface of the earth. However, miners of the sixteenth century lacked much of the knowledge that helps geologists today find ore deposits that are not on the surface.

Locating and Extracting Ores

The modern approach uses knowledge gained from experience. As in Agricola's day, much of the wealth possessed by a mining company is in the form of information regarding the means of best seeking out and retrieving materials from the solid earth. Certain surface geochemical and geophysical indicators help direct the steps of geologists and miners searching for ore. Thus, by the time a company in search of ore begins drilling, a great deal of exploratory work has been done. Only at that point is it possible to determine the value of the deposits, which may simply be minerals of little economic interest.

It is estimated that a cubic mile (1.6 km^3) of average rock contains about $1 trillion worth of metals, which at first sounds promising—until one does the math. A trillion dollars is a lot of money, but 1 cu. mi. (equal to 5,280 × 5,280 × 5,280 ft., or 1,609 km^3) is a lot of space too. The result is that 1 cu. ft. (0.028 m^3) is worth only about $6.79. But that is an average cubic foot in an average cubic mile of rock, and no mining company would even consider attempting to extract metals from an average piece of ground. Rather, viable ore appears only in regions that have been subjected to geologic processes that concentrate metals in such a way that their abundance is usually many hundreds of times greater than it would be on Earth as a whole.

Ore contains other minerals, known as gangue, which are of no economic value but which serve as a telltale sign that ore is to be found in that region. The presence of quartz, for example, may suggest deposits of gold. Ore may appear in igneous, metamorphic, or sedimentary deposits as, well as in hydrothermal fluids. The latter are emanations from igneous rock, in the form of gas or water, which dissolve metals from rocks through which they pass and later deposit the ore in other locations.

Confronting the Hazards of Mining

Mining, a means of extracting not only ores but many industrial minerals and fuels, such as coal, is difficult work fraught with numerous hazards. There are short-term dangers to the miners, such as cave-ins, flooding, or the release of gases in the mines, as well as long-term dangers that include such mining-related diseases as black lung (typically a hazard of coal miners). Then there is the sheer mental and emotional stress that comes from spending eight or more hours a day away from the sunlight, in claustrophobic surroundings.

And, of course, there is the environmental stress created by mining—not just by, the immediate impact of cutting a gash in Earth's surface, which may disrupt ecosystems on the surface, but myriad additional problems, such as the seepage of pollutants into the water table. Abandoned mines present further dangers, including the threat of subsidence, which make these locations unsafe in the long term.

Higher environmental and occupational safety standards, established in the United States during the last third of the twentieth century, have led to changes in the way mining is performed as well as in the way mines are left when the work is completed. For example, mining companies have experimented with the use of chemicals or even bacteria, which can dissolve a metal underground and allow it to be pumped to the surface without the need to create actual underground 'shafts and tunnels or to send human miners to work them.

Industrial Minerals and other Products

Industrial minerals, as noted earlier, are non- metal—containing mineral resources of interest to economic geology. Examples include asbestos, a generic term for a large group of minerals that are highly resistant to heat and flame; boron compounds, which are used for making heat- resistant glass, enamels, and ceramics; phosphates and potassium salts, used in making fertilizers; and sulphur, applied in a range of products, from refrigerants to explosives to purifiers used in the production of sugar.

Just one industrial mineral, corundum (from the oxides class of mineral), can have numerous uses. Extremely hard, corundum in the form of an unconsolidated rock commonly called emery has been used as an abrasive since ancient times. Owing to its very high melting point—even higher than that of iron—corundum also is employed in making alumina; a fire-proof product used in furnaces and fireplaces. Though pure corundum is colourless, trace amounts of certain elements can yield brilliant colours: hence, corundum with traces of chromium becomes a red ruby, while traces of iron, titanium, and other elements yield varieties of sapphire in yellow, green, and violet as well as the familiar blue.

Environmental Impact of Economic Geology

A few decades ago most geologists were engaged in the exploration and development of mineral resources. However, economic geology and the application of geology to problems of the urban environment are slowly demanding the services of an increasing number of geologists. Today a sufficient number of economic geologists are interested in environmental problems (many are not tied to mining, and have interests in geochemistry and petrology). They subscribe to the thought that "mineral resources will always be needed", but "environmental issues are a major factor in mining feasibility".

There is now a greater interest in the environment than ever before and we face concerns about the environmental impacts of almost every aspect of our daily lives. Water, soil, air and the biological environment can all be changed dramatically by the activities of industrial societies like our own, not least through mechanisms which are controlled by essentially geological processes.

Waste disposal, contamination of land by industry, the impacts of mining, water pollution and even air quality (through the dispersion of airborne mineral particles) are affected by geological processes and phenomena controlled by the composition, distribution, structure and behaviour of the underlying rocks. Everyday environmental problems are thus influenced to a greater or lesser extent by geology. The course aims to provide a broad overview of these many and various aspects of environmental geology, providing the scientific framework for understanding key environmental issues.

Environmental concerns are a major factor in determining whether mineral deposits will be developed and mined. Most economic geologists and mining companies support efforts to decrease environmental degradation due to mining.

Environmental Effects

Mining, no less than agriculture, has always been essential to humanity's progress. Indeed, we now make use of most elements in the periodic table. However, as. overpopulation and the quest for a higher standard of living pave increased demand for minerals and metals, concern over the effects of mining and drilling in the natural environment has grown and it has become increasingly apparent that the Earth's resources are not inexhaustible.

In its report in 1987, 'Our Common Future', the United Nations World Commission on Environment and Development pointed out that the world manufactures seven times more goods than it did in 1950. The commission proposed 'sustainable development', a marriage of economy and ecology as the only practical solution, i.e., growth without damage to the environment.

Most mines have an on-site mineral processing plant and many metal mines have a nearby smelter. For a general assessment of the environmental impact of developing new mining operations we must consider the effects of the three. The term mining is here taken to include all extractive operations, e.g., quarrying. The major areas of concerns are dealt with below.

Damage to Land

It has been estimated that the cumulative world use of land for mining between 1976 and 2000 would be about 37,000 square km, that is, about 0.2 per cent of the land surface. More developed countries have a greater proportion of disturbed ground than the less developed. The degree of reclamation of this ground is now accelerating rapidly and good use being made of old holes for the disposal of old mine, domestic and other wastes.

Other mined over areas have been turned into nature reserves and recreational parks. Future mines may be less likely to produce sites for waste disposal as most are now backfilled. This is a very necessary operation as each year an estimated 27,000 Mt of non-fuel minerals and overburden are taken from the earth's crust.

Release of Toxic Substances

Metals are not only important for the usage we make of them, but they are also an integral part of our make-up and that of other living organisms. However, while some metallic elements are essential components of living organisms, either deficiencies or excess of these can be very damaging to life. Excesses in the natural environment can arise when it is penetrated by mine waters which may issue from the mine itself, or from waste heaps.

Some metals, e.g., cadmium, mercury and metalloids, like antimony, arsenic, which are very common in small amounts in many polymetallic sulphides ores and are indeed often recovered as byproducts, are highly toxic, even in small quantities, particularly in a soluble form which can be absorbed by living organisms.

The same applies to lead, but mercifully it is fairly unreactive unless ingested and fortunately most lead minerals that form in nature are very insoluble in ground water. Cyanide has long been used for gold extraction in mineral processing plants and in the world's largest gold field, the Witwaterstrand Basin, USA, there is major contamination of surface water with cobalt, manganese, lead, and zinc as a result of the cyanidation process and oxidation by acid mine waters. Cyanide itself is not a problem as it breaks down under the influence of ultraviolet light in near surface layers. Nevertheless, in developed countries, legislation now requires the establishment of cyanide neutralisation plants at all industrial undertakings using this chemical.

Acid Mine Drainage

Acid waters generated by present or past mining result froth the oxidation, in the presence of air, water and bacteria, of sulphide minerals, particularly pyrite. They may therefore develop in coal fields as well as orefields. Suphuric acids and iron oxides are generated. The acid attacks other minerals, producing solutions which may carry toxic elements, e.g., cadmium, arsenic into local environment. Acid water generation may occur during the exploration, operation and closure stages of a mine. These waters may issue from three main sources: the mine's dewatering system; tailings disposal facilities; and water heaps.

The discharge may only produce minor effects such as local discolouration of soils and streams with precipitated iron oxides, or lead to extensive aerial pollution of whole river system and farmland. In some mining fields, this problem is at its worst after mine closures have taken place. This is due to the water table rebound that occurs after pumping equipment is removed and this has become an urgent problem in British coalfields, which were and are dominantly underground mines working high sulphur coals, as mine closures accelerated over the last decade.

Industrial minerals Industrial mineral operations have the same general environmental impact on land and ground water disturbances as metalliferous or coal-mining, although the impact is generally less marked since the mines are usually smaller and shallower, and normally less waste is produced because in most cases ore grades are higher than in metal mining.

Pollution hazards owing to heavy metals or acid waters are low or non-existent and atmospheric pollution, caused by the burning of coal or the smelting of metallic ores, is much less serious or absent. The excavations created by industrial mineral operations are often close to conurbations, in which case these holes in the ground may be of great value as landfill sites for city waste.

Legal Measures

Legal means of enforcing anti-pollution measures are very necessary, although it must be pointed out that many international mining companies are now adhering to the strictest self-regulation even in countries where such legislation is minor or non-existent.

Environmental Impact Statements

In many countries it is now mandatory that a company proposing to apply for planning permission to commence a mineral operation prepares such a statement. This covers every aspect from effects on vegetation, climate, air quality, noise, ground and surface water to the proposed methods of ground reclamation at the termination of the operation. In some countries a bond must be deposited to assure that reclamation does take place.

These statements must include records of the condition of the environment in the potential mining area when planning permission is applied for. Companies now collect such data during exploration stage, including surface descriptions and photographs, geochemical analyses showing background levels of metals and acidity and details of the flora and fauna.

From the planning and regulatory authorities' point of view these reports present the most effective means of minimising detrimental effects from the outset, but they may also be of great benefit to the developer because (i) they will help to obtain planning permission in the shortest possible time, and (ii) they often reveal aspects of the operation requiring attention at the outset and thus avoid expensive modifications in the future.

Bugs and in Situ Mining

Many sulphide deposits, i.e., porphyry coppers, are overlain by oxidised ores. Such ores can be mined, if necessary, by fracturing them by blasting and then pumping acid solutions through the rock to dissolve metals such as copper and uranium. The metal-bearing solutions are pumped to the surface and the metals recovered. Very low grade, small and otherwise economically non-viable deposits can be exploited and the process can be used at considerable depths.

Outlook

Measures such as recycling and substitution and new materials technology will play their part in reducing the impact of mineral exploitation on the environment, but in the immediate future we have to look to an increasing sense of responsibility by all those involved in any way with the industry, be they developers or regulators.

There are many hopeful signs that this is taking place: for example, in 1992, 19 major mining corporations from five continents joined together to form the international council on metals and the environment whose remit is to promote the development, implementation, and harmonisation of sound environmental and health policies and practices which will ensure the safe production, use, recycling and disposal of metals.

Engineering Geology

Engineering geology is the application of the geology to engineering study for the purpose of assuring that the geological factors regarding the location, design, construction, operation and maintenance of engineering works are recognized and accounted for. Engineering geologists provide geological and geotechnical recommendations, analysis, and design associated with human development and various types of structures. The realm of the engineering geologist is essentially in the area of earth-structure interactions, or investigation of how the earth or earth processes impact human made structures and human activities.

Engineering geology studies may be performed during the planning, environmental impact analysis, civil or structural engineering design, value engineering and construction phases of public and private works projects, and during post-construction and forensic phases of projects. Works completed by engineering geologists include; geological hazard assessments, geotechnical, material properties, landslide and slope stability, erosion, flooding, dewatering, and seismic investigations, etc. Engineering geology studies are performed by a geologist or engineering geologist that is educated, trained and has obtained experience related to the recognition and interpretation of natural processes, the understanding of how these processes impact human made structures (and vice versa), and knowledge of methods by which to mitigate against hazards resulting from adverse natural or human made conditions. The principal objective of the engineering geologist is the protection of life and property against damage caused by various geological conditions.

An engineering geologist logging rock core in the field, Western Australia.

The practice of engineering geology is also very closely related to the practice of geological engineering and geotechnical engineering. If there is a difference in the content of the disciplines, it mainly lies in the training or experience of the practitioner.

Training of Geologist

One of the most important roles as an engineering geologist is the interpretation of landforms and earth processes to identify potential geologic and related man-made hazards that may have a great impact on civil structures and human development. The background in geology provides the engineering geologist with an understanding of how the earth works, which is crucial minimizing earth related hazards. Most engineering geologists also have graduate degrees where they have gained specialized education and training in soil mechanics, rock mechanics, geotechnics, groundwater, hydrology, and civil design. These two aspects of the engineering geologists' education provide them with a unique ability to understand and mitigate for hazards associated with earth-structure interactions.

Scope of Studies

Engineering geology investigation and studies may be performed:

- For residential, commercial and industrial developments;
- For governmental and military installations;
- For public works such as a stormwater drainage system, power plant, wind turbine, transmission line, sewage treatmentplant, water treatment plant, pipeline (aqueduct, sewer, outfall), tunnel, trenchless construction, canal, dam, reservoir, building foundation, railroad, transit, highway, bridge, seismic retrofit, power generation facility, airport and park;

- For mine and quarry developments, mine tailing dam, mine reclamation and mine tunneling;
- For wetland and habitat restoration programs;
- For government, commercial, or industrial hazardous waste remediation sites;
- For coastal engineering, sand replenishment, bluff or sea cliff stability, harbor, pier and waterfront development;
- For offshore outfall, drilling platform and sub-sea pipeline, sub-sea cable; and
- For other types of facilities.

Geohazards and Adverse Geological Conditions

Typical geologic hazards or other adverse conditions evaluated and mitigated by an engineering geologist include:

- Fault rupture on seismically active faults;
- Seismic and earthquake hazards (ground shaking, liquefaction, lurching, lateral spreading, tsunami and seiche events);
- Landslide, mudflow, rockfall, debris flow, and avalanche hazards;
- Unstable slopes and slope stability;
- Erosion;
- Slaking and heave of geologic formations, such as frost heaving;
- Ground subsidence (such as due to ground water withdrawal, sinkhole collapse, cave collapse, decomposition of organic soils, and tectonic movement);
- Volcanic hazards (volcanic eruptions, hot springs, pyroclastic flows, debris flow, debris avalanche, gas emissions, volcanic earthquakes);
- Non-rippable or marginally rippable rock requiring heavy ripping or blasting;
- Weak and collapsible soils, foundation bearing failures;
- Shallow ground water/seepage;
- Other types of geologic constraints.

An engineering geologist or geophysicist may be called upon to evaluate the excavatability (i.e. rippability) of earth (rock) materials to assess the need for pre-blasting during earthwork construction, as well as associated impacts due to vibrationduring blasting on projects.

Soil and Rock Mechanics

Soil mechanics is a discipline that applies principles of engineering mechanics, e.g. kinematics, dynamics, fluid mechanics, and mechanics of material, to predict the mechanical behaviour of soils. Rock mechanics is the theoretical and applied science of the mechanical behaviour of rock and rock masses; it is that branch of mechanics concerned with the response of rock and rock masses to the force-fields of their physical environment. The fundamental processes are all related to the behaviour of porous media. Together, soil and rock mechanics are the basis for solving many engineering geology problems.

Methods and Reporting

The methods used by engineering geologists in their studies include:

- Geologic field mapping of geologic structures, geologic formations, soil units and hazards;
- The review of geologic literature, geologic maps, geotechnical reports, engineering plans, environmental reports, stereoscopic aerial photographs, remote sensing data, Global Positioning System (GPS) data, topographic maps and satellite imagery;
- The excavation, sampling and logging of earth/rock materials in drilled borings, backhoe test pits and trenches, fault trenching, and bulldozer pits;
- Geophysical surveys (such as seismic refraction traverses, resistivity surveys, ground penetrating radar (GPR) surveys, magnetometer surveys, electromagnetic surveys, high-resolution sub-bottom profiling, and other geophysical methods);
- Deformation monitoring as the systematic measurement and tracking of the alteration in the shape or dimensions of an object as a result of the application of stress to it manually or with an automatic deformation monitoring system.

The fieldwork is typically culminated in analysis of the data and the preparation of an engineering geologic report, geotechnical report or design brief, fault hazard or seismic hazard report, geophysical report, ground water resource report or hydrogeologic report. The engineering geology report can also be prepared in conjunction with a geotechnical report, but commonly provides the same geotechnical analysis and design recommendations that would be presented in a geotechnical report. An engineering geology report describes the objectives, methodology, references cited, tests performed, findings and recommendations for development and detailed design of engineering works. Engineering geologists also provide geologic data on topographic maps, aerial photographs, geologic maps, Geographic Information System (GIS) maps, or other map bases.

Environmental Geology

Environmental geology is a young interdisciplinary science which bring about, besides of new approaches to solving problems of connections between the geoenvironment and the man, new concepts and terms or enhancement of old concepts with new aspects. The most significant of them are: geoenvironment, geological factors of the environment and environmental geologic conditions. The principal role of environmental geology in the geoenvironment management is to ensure its rational use and protection.

The geoenvironment on the one hand affords the human community to take advantages of the mineral and groundwater resources and other favourable geologic phenomena, necessary for its existence and evolution, on the other hand it endangers the mankind by harmful geologic phenomena and brings about unfavourable conditions for the development of the society. Both types of these geofactors form together environmental geologic conditions of a territory, which bring the man benefits or problems in his evolution. Solving these matters, leading to the geoenvironment rational use and protection, is the main role of environmental geology in management of the geoenvironment.

The term environment has been used about forty years. The environment is that part of universe, which is in interaction with the man, which the man exploits and affects and to which he adapts. Some different definitions appeared later taking into account, in contrast to this in substance natural scientific definition, as well as the socio-economic relations.

In connection with the original definition of the environment the geological environment (geoenvironment) can be defined as that part of lithosphere, which directly influences the conditions of the existence and development of the society, which the man exploits and converts. The basic components of the geoenvironment (geocomponents) are rocks, relief, and groundwater and geodynamic phenomena.

This meaning of the concept has been fixated only recently, in connection with the creating and using of the term and concept the environment. Until lately the term geoenvironment was used as the synonym of the lithosphere or the earth crust, i.e. in the meaning of the study object of geology, or in the meaning of a matter, i.e. as the space fill with rocks including pores and cracks fill with water. In the last meaning the term is used sometimes at present as well ñ to differentiate the concept of the rock environment, which does not include the free or non-adhesive water. The depth range of the geoenvironment depends on the scope of human interference with the earth crust, i.e. about 10 km at present.

Environmental-geologic Conditions and Geofactors of the Environment

Upon the geoenvironment can be introduced the system of various branches of geology: engineering geology, hydrogeology, economic geology, etc. Each branch defines therein geocomponents from the standpoint of its specialization. Geocomponents defined from standpoint of engineering geology, for instance, create the system of engineering-geological conditions. Relevant properties of the geocomponents in this system can be the factors of their other properties or the engineering-geological factors of some mansë activities (e.g. of construction, opening and exploitation of stone pits or sand pits, etc.). In the same way we can define hydrogeological, geophysical and others conditions and factors.

A lot of these way defined factors are of specific meaning and affect the objects and relations that merely the corresponding science is dealing with. However, some of them have effects on the quality of the environment and on land use, too. These are integrated into environmental-geologic conditions (geoenvironmental conditions), which are defined as the complex of geocomponets properties affecting the quality and the way of geoenvironment and landscape exploitation. These properties of the geoenvironment we call as the environmental-geologic factors or as geological factors (geofactors) of the environment.

Hence it follows, that we have specific properties of geocomponets, which can be the factors of their other properties, the geofactors of specific human activities (e.g. of construction) and the geofactors of the environment. The latter ones represent the usable properties of the geoenvironment such as available resources of mineral raw materials, suitability for construction, etc. But, besides of geofactors that provide possibilities of the geoenvironment exploitation, which are called as geological potentials (geopotentials) of the environment, there are another geofactors, called as geological barriers (geobarriers) of the environment, which limit or in a negative way affect the land-use. To this group of geofactors belong geological hazards (geohazards), which endanger the environment or even health and life of people (ruinous earthquakes, catastrophic landslides, etc.) and geological constraints (geoconstraints), which limit the land use and/or raise the price of land-use (unsuitable foundation soils, susceptibility to landslides, etc.). The complex of all geofactors in a territory creates its environmental geologic conditions.

Table: Geological factors of the environment and environmental-geologic conditions of a territory.

<table>
<tr><td>Geological potentials</td><td>Suitable condition for construction</td><td>Available resources of groundwaters</td><td>Available resources of mineral raw materials</td><td>Phenomena useful in health service, tourism and recreation</td><td rowspan="2">Environmental geologic condition of the territory</td></tr>
<tr><td>Geological barriers</td><td>Landslides
Earth flows
Avalanches</td><td>Pollution of groundwaters</td><td>Subsidence and sagging of the surface</td><td>Sheet erosion
Wash out erosion</td></tr>
</table>

The best way how to express environmental-geologic conditions of a territory is to compile a map. In Slovakia, there are geoenvironmental conditions usually depicted in maps of geofactors of the environment. The maps consist of several map sheets covering practically environmental aspects of all branches of geology mainly engineering geology, hydrogeology, geochemistry, economic geology, geophysics and pedology. Regarding the great amount of the maps, as well as the high degree of specialisation of some of them, a simplify comprehensive Map of significant geofactors is compiled for purposes of land-use planning and environmental protection. This map contains only some geofactors (taken over from the above maps) which in a high degree affect the land-use: significant mineral and water resources, unsuitable foundation soils, high quality fertile soils, seismic intensity, important landslides, high radon risk districts, and some others. An example of such map was published in Proceedings from 8th IAEG Congress in Vancouver. However this map does not contain all geofactors needful for representation of geoenvironmental conditions of a territory (contaminated soils and waters, sensitive or vulnerable rocks and soils, etc.). Thus the question how a proper map of geoenvironmental conditions has to look like is still waiting for an answer.

The Role of Environmental Geology in the Geoenvironment Management

The man in the course of his evolution employed all the time in a greater extent geopotentials, struggled against geohazards and surmounted geoconstraints taking always more ingenious measurements. But sometimes, in course of various activities, he brought about geohazards or deteriorated the geoenvironment by himself. Hence it follows that the main role of environmental geology in the geoenvironment management is the care of its rational use and protection.

The concept rational use of the geoenvironment covers generally these aspects:

1. Most useful from point of view of economic demands.
2. Without harmful impacts on geoenvironment and landscape.

3. Technical and economic unpretending.
4. Architectonic and landscape suitable.

The geoenvironmental conditions of a territory enable generally various possibilities of land use (mining of mineral raw materials, civil and industrial construction, transport construction, ground water exploitation, recreational, sanitary and sporting purposes, etc.). Decision making process in that case is aimed at the best way of land use from point of view of nation-wide or regional economic requirements. So, the first demand covers possible collisions of interests in a territory.

The second point tends care on trouble free exploitation of the geoenvironment and landscape, i.e. without unacceptable consequences such as invocation or activation of harmful geodynamic phenomena (landslides, suffosion, erosion a. o.), pollution of groundwater and rock environment, negative impact on superficial water and biota, etc.

Technical and economic unpretending exploitation of the geoenvironment requires the choice of reasonable methods of mining or constructions foundation and optimum sitting of activities in landscape. That calls for sites, which do not necessitate protection against stability failure or against pollution of the geoenvironment, alternatively only unassuming and inexpensive ones.

The last point calls for designs in harmony with nature. It means that interventions into the geoenvironment and landscape must be technically and ecologic sound and aesthetic suitable.

From listed above results that the concept rational use of the geoenvironment includes also its protection. Why than we speak separately of the geoenvironment protection? It is mainly in connection with old objects of techno-sphere, which were not built up in accordance with requirements of the geoenvironment rational use, in case of old ecologic loads, on the chance of accidents and in case of unstable territories (slope movements, erosion, subsidence, sagging a. o.). In general we distinguish the geoenvironment protection against pollution and against stability, integrity and structure failure, which bring about instability of a territory.

In the case of the protection against pollution we basically use four methods, geoenvironment injury. Some of them are listed below:

1. Removal of the techno sphere objects (constructions, embankments),
2. Removal of unstable parts of rocks mass (blocks of rocks, landslides),
3. Measures against slopes stability failure (abutment walls, anchoring),
4. Anti-erosion measurements (modification of slopes, method of farming),
5. Countermeasure against sagging, suffosion, liquefaction,
6. Monitoring of unstable territories.

Solution of these problems needs an extensive interdisciplinary approach. Consequently environmental geology is an interdisciplinary science, as within the system of geological sciences as in its position at the boundary of geology and other natural sciences, as well as technical and socio-economic sciences. It analyses all environmental aspects of the geoenvironment taking into account mainly its evaluations from point of view of engineering geology, hydrogeology, economic geology, geochemistry and geophysics. At the same time it utilizes the knowledge of other geological sciences as physical geology, tectonics, mineralogy, petrology, etc. Knowledge of technical sciences, mainly of geotechnics it applies at proposals of rational means of the geoenvironment exploitation, at solving of problems of manís interference with geoenvironment and at proposals of the environment protection from harmful geological processes and from pollution.

Planetary Geology

Planetary geology is a branch of geology devoted to the study of structure, composition, processes, and origin of major and minor planetary bodies traditionally inside and now, in the twenty-first century, outside the solar system, and to the effects of interaction between planetary bodies within the solar system. Planetary geology interfaces with many other fields including mathematics, astronomy, biology, chemistry, and physics. Planetary geologists work in the field, laboratory, and—indirectly—in outer space through imagery and other data returned by spacecraft. One planetary geologist has visited another planetary body and perhaps someday in the near future other scientific explorers will visit the moon, Mars, and other planetary bodies.

The goal of planetary geology studies of other planetary bodies is to understand the origin of the features seen (and samples returned, if such are available). Planetary geology studies usually relate to the quest for an understanding of the geological history of the body from its formation during accretion from the early solar nebula to its present condition. Most planetary bodies in the solar system of any significant size have passed through stages of accretion, internal heating and differentiation, and surficial and tectonic evolution. However, some objects have not, for example, small asteroids, comets, and smaller solar system debris such as meteorites. These objects can give important clues to the nature of the origin of the solar system, as they contain unaltered primordial material from the original nebular dust cloud. An important sub-discipline within planetary geology is meteoritics, the study of meteorites.

Most initial planetary geology studies help establish a temporal (time) framework for major events in the history of a planetary body. This framework, called a geologic time scale, is used to help fit events into a sequential order. On Earth, this kind of framework was established over a period of about 200 years using fossils, age relations among rocks, and absolute (radiometric) age dating. For planetary bodies like the moon and Mars, for which scientists have some surface samples (there are a few Martian and lunar meteorites plus the samples returned by lunar landing missions), some radiometric age dating is possible. Mainly, planetary age relations are established by examining photographic evidence such as impact-crater densities, cross-cutting relationships, and superposition relationships (layers lying on top of one another). This is worked out by careful geologic mapping using photographic mosaics of planetary surfaces.

Planetary geologists are also interested in studying interactions among planetary bodies. When one planetary body impacts another, great energy can be released. Large body impacts represent the most energetic geologic process known. Very large body impacts can release the energy equivalent to thousands or even millions of megatons of TNT (megaton = 1 million tons) in a few seconds. Impact events produce impact craters, whose size and shape are controlled by factors like planetary gravity, strength of crustal materials, and density of planetary atmosphere. The impact crater is the most common feature on rocky and icy planetary surfaces in the solar system, except where resurfacing processes (like volcanism, erosion and sedimentation, and crustal melting) remove impact craters faster than they can form.

Studies of large impact craters on Earth have lead geologists to new understandings of how large body impacts, and the ecological disasters that they can cause, have affected evolution of life on Earth. Some large body impact events have been strongly connected to mass extinctions of life on Earth, a driving force in evolutionary change. The best modern example of this is the great Cretaceous-Tertiary mass extinction event (65 million years ago), which has been dated as the same age as two large impact craters on the Earth (Chicxulub crater in Mexico, nearly 112 mi [180 km] in diameter, and Boltysh crater in the Ukraine, about 15 mi [24 km] in diameter). Planetary geologists are currently working on theories about possible comet swarms due to perturbations in the Oort cloud (the region about halfway between the sun and the helio-pause) or other disruptive mechanisms which may have made time intervals in the past more dangerous for life on Earth due to cosmic impacts.

Studies of impact craters hold considerable academic interest for investigators, but on the practical side, considerable natural resources have been discovered within impact craters. For this reason, their study has been supported by petroleum, mining, and ground-water concerns. The intensive fracturing caused by impact creates avenues for fluid movement within the affected crustal area, causing entrapment of hydrocarbons and emplacement of some types of fluid-borne mineral resources. Groundwater accessibility and quality is either enhanced or degraded in impact areas, depending upon geological conditions.

Studies of impacts of large comets may have other implications for solar system evolution, as recent theories suggest such impacts may have delivered significant amounts of water and primitive organic material to Earth and Mars during a very early phase of their evolution. Recent investigations of the origin of Earth's moon suggest that a major impact event that penetrated Earth's crust ejected material into orbit that eventually became the moon. This theory helps explain some of the unusual chemical characteristics of lunar rocks and their similarity to Earth's upper mantle.

As of October 2006, National Aeronautics and Space Administration's (NASA's) Mars Reconnaissance Orbiter (MRO), which was launched on August 12, 2005, is expected to begin its primary mission in November 2006. It will conduct exploration and reconnaissance of Mars' landforms, minerals, water/ice, stratigraphy (rocks) while in orbit about the planet. Its explorations will help to pave the way for future manned and unmanned missions to the planet.

In 2007, the Kepler spacecraft, a space telescope especially designed to scan large areas of the sky for transits by planets as small as Earth, will be launched by the U.S. National Aeronautics and Space Administration (NASA). By 2011, Kepler should have gathered enough data to pinpoint hundreds of extrasolar planets and to determine how typical the own solar system is in the universe. This is of interest to scientists because estimates of the probability that life exists elsewhere in the universe depend strongly on the existence of planets, and their physical geology, not too different from Earth. Intelligent life is unlikely to evolve on large gas giants or on bodies of any type that orbit very near to their stars or follow highly eccentric orbits. If solar systems like Earth's are rare in the universe, then life (intelligent or otherwise) may be correspondingly rare. Theoretical models of the formation of solar systems have been in a state of rapid change under the pressure of the rush of extra-solar planet discoveries, and revised models indicate that solar systems like Earth's may be abundant. However, these models supply only educated guesses, and must be checked against observation.

Consequently, in the 2000s, planetary geology is rapidly expanding into the study of extra-solar system bodies. This future area for planetary geology investigation will help shed more light on the origin of the own solar system and on solar system formation in this part of the Milky Way galaxy.

Geomicrobiology

Geomicrobiology is the interdisciplinary study of the interactions of microorganisms with earth materials. It concerns the role of microbes in geological and geochemical processes. The field of geomicrobiology has revealed new insights into the intersection of life with the physical and chemical composition of Earth's surface. Soil microbes play a large role in the transformations of elements and minerals. The interactions between microbes and elements and minerals is especially important in the near surface environment known as Earth's Critical Zone, where biotic and abiotic factors regulate the conditions for life-sustaining resources. Virtually all elements can be transformed by microbes and many elemental cycles depend on soil microbes. The relationships between microorganisms and these inorganic compounds have important implications for both the surrounding natural environment and human use. Soil microbes can be used by humans in mineral resource extraction and bioremediation.

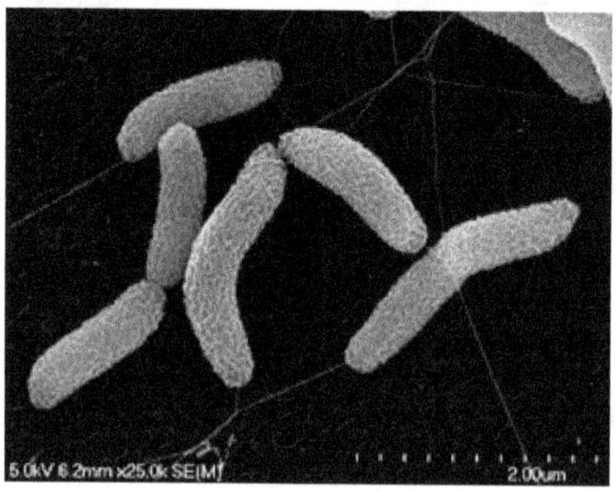

Scanning electron micrograph of Desulfovibrio desulfricans, a Gram-negative sulfate-reducing bacteria.

Key Processes of Geomicrobiology

Geomicrobiological processes are relevant in many natural environments including aquifers, geological and geochemical processes, extreme environments (acidic, extreme temperatures and saline conditions) and metal ion reduction.

Changes in the mobility and availability of metal is governed by redox reactions, complexation by metabolites and siderophores, and methylation. Immobilization transforms metals and other inorganic substances into microbial biomass. Oxidation and reduction reactions are used by microbes to gain energy during respiration and photosynthesis, resulting in the immobilization and mineralization of metals and nutrients. Redox reactions also drive the nitrogen cycle, and reduce toxic compounds. Conjugation reactions are another process by which toxic chemicals are converted to polar components. These conjugated substances are then made unavailable by immobilization or they become recalcitrant and accumulate in the food web.

Weathering

Weathering plays an important role on Earth ecosystems and is important for the release of nutrients into the biosphere as a result of rock dissolution or the regulation of long-term climate by the consumption of atmospheric carbon dioxide from silicate alterations. Weathering has been considered a strictly physical and chemical process. Nonetheless, many weathering processes can be affected by the presence of microbial communities. All types of rocks are susceptible to microbial weathering, including siliceous and calcareous rocks. In general microbial weathering is due to the formation of organic acids or the production of metal-chelating siderophores on the surface of rock or minerals. Other processes include oxidative or reductive conditions of metals on rocks or minerals.

Precipitation of Carbonates and Phosphates

The influence of bacteria in mineral precipitation has been described in different natural habitats including aquatic environments and terrestrial systems. Carbonate precipitation is controlled by bacterial processes and one of the mechanisms proposed is the oxidation of ammonium by ammonium oxidizers which decreases environment pH. Other authors report that the matrix of extracellular polymeric secretions (EPS) affects mineral precipitation Simultaneous precipitation of struvite (MgNH4PO4•6H2O) associated with bacterial activity has been described in vitro carbonate precipitation experiments as well as carbonate and phosphate precipitation by bacteria from saline soils.

Examples of Microbial Metabolism of Minerals and Inorganic Substances

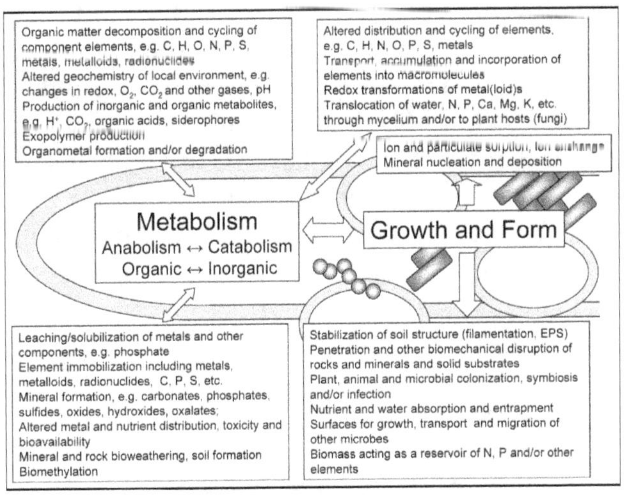

A comprehensive depiction of the effects of microbial mineral metabolism.

Table of Microbial Roles for Representative Elements	
Elements	Microbial roles in elemental cycles
Phosphorus (P)	Dissolution of inorganic phosphate minerals in soils; Decomposition of organic phosphorus; Release of organically bound phosphorus by use of phosphatase enzymes; Assimilation and transformation of inorganic phosphorus species; Transformations of soil organic phosphorus; Phosphorus transfer to plants by mycorrhizae.

Sulfur (S)	Degradation of organic sulfur compounds; Sulfur transformations; Assimilation of organic and inorganic sulfur; Sulfidogenesis, Sulfur accumulation, SO_4^{2-} reduction and assimilation; Oxidation of H_2S to S(o) and reduction of S(o) to H_2S.
Iron (Fe)	Bioweathering of iron-containing minerals in soil; Iron solubilization by siderophores; Fe(III) reduction to Fe(II); Fe(II) oxidation to Fe(III); Iron biomineralization to oxides, hydroxides, carbonates, sulfides, etc. Metal sorption to iron oxides.
Manganese (Mn)	Mn(II) oxidation and immobilization as Mn(IV) oxides; Bioaccumulation of Manganese oxides to surfaces and exopolymers; contribution to desert varnish formation; biosorption; accumulation; intracellular precipitation; biomineralization; metal sorption to manganese oxides.
Magnesium (Mg), Calcium (Ca), Nickel (Ni), Zinc (Zn), Cadmium (Cd), and Strontium (Sr)	Bioweathering of minerals in soil; biosorption; uptake and accumulation; bioprecipitation of oxalates, sulfides, carbonate, phosphates, etc.
Copper (Cu)	Mobilization from copper containing minerals in soils; CuS formation; biosorption; uptake and accumulation; bioprecipitation.
Mercury (Hg)	Biomethylation; Reduction of Hg(III) to Hg(o); Oxidation of Hg(o) to Hg(II). Mercury volatilization as Hg(o); degradation of organomercurials; biosorption; accumulation.
Selenium (Se)	Reductive transformation of Se (VI) to Se(IV) to Se(o); Se(o) oxidation; Biomethylation and demethylation; assimilation of organic and inorganic Se compounds.

Manganese

In soil, Mn typically exists as bound to metal or as an oxide, and is most abundant as MnO_2. Of the several oxidation states Mn has in nature, only three (II, III, and IV) are biologically important. Other than as a component of enzymes in living organisms, Mn also serves as a source of energy (i.e. as an electron acceptor and donor). Furthermore, of the three biologically important oxidation states, only Mn (II) is soluble, Mn-reduction may also serve to extract mineral Mn from soil to fulfill nutritional needs Mn(II)-oxidizing bacteria are a diverse group found in almost all environments. These bacteria are up to 5 orders of magnitude faster than abiotic reactions in the production of Mn oxides which have an amorphous structure with a high surface area. The mechanism of Mn(II) oxidation by these bacteria is not clear, although recent outcomes from studies with Bacillus sp. strain SG-1 have shown that a Mn(III) is an intermediate in the final oxidation of Mn(II) through enzymatic activity.

Uranium-nitrate Relationship

Uranium occurs naturally in soils, and is mined for use in energy and weapon production. Contamination of aquifers by Uranium produced from nuclear weapons and fuel is becoming a real problem in different parts of the world. These aquifers are generally oxidized, therefore uranium predominantly exists in the reactive state U(VI). The U(VI) is often coupled with carbonate making the compound quite soluble and easy to leach. Uranium reduction occurs under anaerobic conditions with Fe(III) and/or sulfate reduction. The biogeochemical processes that occur after uranium reduction are poorly understood. Recent research has shown that the addition of nitrate (a common co-contaminant with uranium) to U(IV)-containing sediments leads to the oxidation and remobilization of U(IV). Microorganisms responsible in this process are believed to be associated with nitrate dependent Fe(II)-oxidizing microorganisms.

Mineral Cycles involving Microbes

Microbes are involved in and drive many geochemical cycles. Among them, the nitrogen cycle, the phosphorus cycle, the sulfur cycle, and the iron cycle.

Phosphorus Cycle

Phosphorus is an important nutrient for all living organisms. Most phosphorus found in living systems are in the form of inorganic phosphate. On Earth virtually all known phosphorus exists in the +5 oxidation state. Nonetheless, there are also two additional known forms: phosphonates (+3) and phosphonates (+1). Microbes immobilize, mineralize, and solubilize phosphorus.

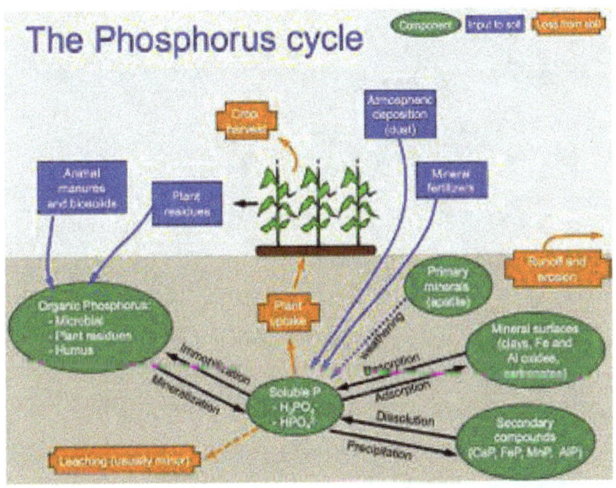

The Phosphorus Cycle, from a soil perspective.

The phosphorus cycle does not include a gas phase, and thus very little phosphorus makes its way into the atmosphere (small amounts of phosphoric acid can make their way into the atmosphere to contribute to acid rain). The phosphorus cycle begins in sedimentary rock and is released into the soil and water through weathering. Microbes (and plants, as well) assist in the solubilization of mineral phosphorus in soil through the production of acids to lower the soil pH, as well as the secretion of chelating agents to bind the metal ions that would bind to the phosphate. These soluble phosphates are immobilized by living organisms, and can be produced through mineralization reactions.

Actinomycetes, Pseudomonas, and Bacillus species are among the bacteria that solubilize inorganic phosphorus. Aspergillus and Penicillium are among the fungi that solubilize phosphorus. Additionally in the rhizosphere, mycorrhizal symbioses between fungi and plants greatly increase phosphorus uptake by plants. Some of the fungi involved are Zygomycetes in the order of Mucorales and Ascomycetes.

Microbes also increase the mineralization process of phosphorous in soil by using phosphatase enzymes to hydrolyze organic phosphorus and convert it back to its inorganic form.

Sulfur Cycle

Most sulfur is contained in rock, with small amounts in the atmosphere and water. In soil, however, most sulfur is organic, as part of an organism; the addition of new sulfur into the ecosystem is through the weathering of sulfur-containing stone. Thus, the sulfur cycle in soil is driven by microorganisms. Some of the microbially-induced transformations sulfur undergoes are: reduction and oxidation, mineralization and immobilization, and volatilization. In short, sulfur bacteria convert mineral sulfur into soluble forms, which are then taken up into microbial biomass, or converted into other mineral forms. Sulfur exits the soil through leaching or volatilization of sulfur.

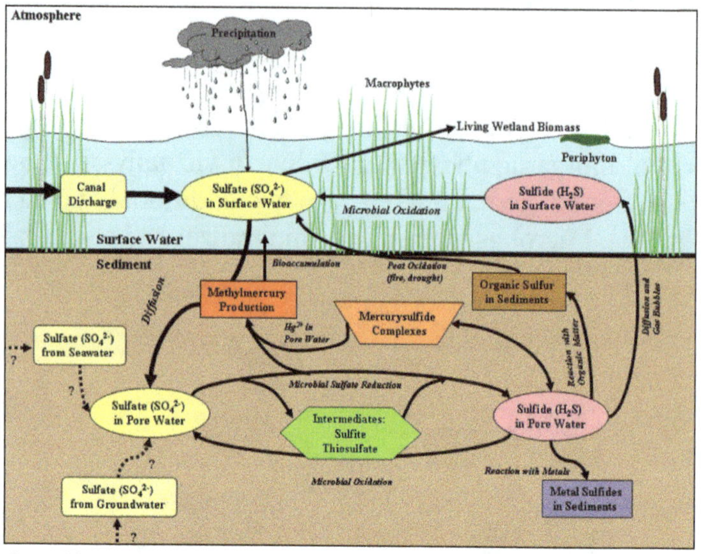

The sulfur cycle: This example shows sulfur cycling in the Everglades.

As an elemental substance, sulfur is important in that it has a wide range of stable redox states. Inorganic sulfur can be used as a terminal electron acceptor for energy. Sulfate-reducing bacteria are comprised of several groups of bacteria that use sulfate as an oxidizing agent, reducing it to sulfide. Most sulfate-reducing bacteria can also use other oxidized sulfur compounds such as sulfite and thiosulfate, or elemental sulfur. This type of metabolism is called dissimilation, since sulfur is not incorporated or assimilated into any organic compounds.

Anaerobic sulfur reducing bacteria reduce sulfate ions to hydrogen sulfide. These sulfur reducing bacteria are heterotrophic organisms that use sulfur ions as terminal electron acceptors in their metabolism. These organisms include bacteria of the genera Desulfovibrio and Desulfotomaculum.

The bacterial genera Thiobacillus, Sulfolobus, and Thiomicrospira oxidize sulfur to produce sulfuric acid as a byproduct. These bacteria accelerate the generation of acid rock drainage (ARD) from pyritic and pyrrhotitic rocks. Acidithiobacillus thiooxidans also produces sulfuric acid and, in conjunction with others of the same genus, is currently used in a mining technique called bioleaching whereby metals are extracted from their ores through oxidation. The bacteria are used as catalysts in the oxidation reaction. Sulfate-reducing bacteria have been considered for remediating water contamination produced by other bacteria from acid mine tailings.

Iron Cycle

After oxygen, iron is the most abundant redox-active element in the Earth's crust. However, it is not easily available in its preferred state. The availability of iron depends on the aerobic condition and pH level of the soil. Under aerobic conditions, iron readily binds with oxygen, forming insoluble (and therefore non-bioavailable) iron oxides. However, at low pH, both common forms of Fe (II, III) are readily soluble. In terms of mineralization and solubilization, microbes accomplish both through redox reactions . Iron can be mineralized, immobilized, oxidized, and reduced.

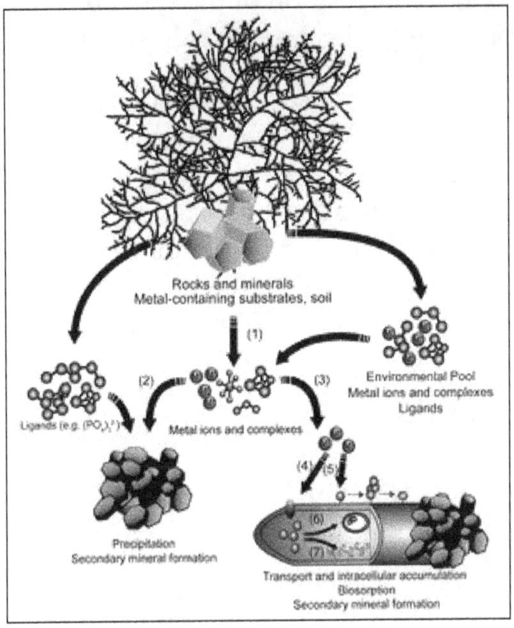

A representation of the general metal cycle. The iron cycle, in specific, is known as the ferrous wheel - a cycle where Fe(III) is microbially reduced into Fe(II), which is then chemically or microbially oxidized back into Fe(III). The reduction of Fe(III) occurs through anaerobic respiration, where iron is used as a terminal electron acceptor. The oxidation of Fe(II) occurs through several pathways, and can occur in both aerobic and anaerobic conditions.

Fe(III) is an important electron acceptor in organic matter decomposition in saturated soil. As this form of iron is soluble, the iron leaves the soil; this causes gleying of the soil (the color of the soil becomes grey or blue), as the color of the sand's sand content becomes dominant. Fe(III) reduction also corrodes steel which causes many industrial problems. One of the most common iron redox reactions in anaerobic soils is the mineralization of iron through binding sulfur to iron, producing pyrite. Furthermore, in these flooded conditions, iron is also used as an energy source; anaerobic iron-reducing bacteria use iron as an terminal electron acceptor in respiration.

Siderophores are iron transporting ligands that are produced by bacteria and fungi that reduce Fe(III) to Fe(II). When microbes experience iron deficiency, siderophores are secreted into the environment to capture iron. The solubilized iron is then taken up by the microbe through a siderophore receptor protein on the outer cell membrane. Iron is the only known essential element for which these specific organic shuttles operate.

Examples of Geomicrobiology interactions in Nature

Rhizosphere

In terms of geobiology, microbes in the rhizosphere serve to alter soil concentration of metals and other minerals In terms of geobiology, microbes in the rhizosphere serve to alter soil concentration of metals and other minerals to assist plants in a symbiotic manner, both indirectly and directly. Low concentrations of bioavailable minerals activate and stimulate the activity of these microorganisms in these biogeochemical processes. Microbes capable of generating these processes, altering the mineral concentration and composition of the rhizosphere soil, include those part of the nitrogen cycle, siderophore-producing microbes, and phosphorus-solubilizing microorganisms.

Iron, as an element, is relatively abundant in soil; however, most of that iron is sequestered in various inorganic iron compounds inaccessible to both plants and microbes. Iron deficiency stimulates the production and secretion of siderophores. Microbial siderophores are chelating agents (organic compounds that can bind to metal ions) secreted by bacteria and fungi, used to capture free iron ions in the soil and, as highly soluble substances, transport that iron to the microorganisms. More importantly from a geochemical point of view, siderophores can also increase the concentration of iron ions in the soil through stimulating the release of inorganic iron (usually from iron oxide crystals, which trap most of the inaccessible iron in the soil). There are three general categories of methods used to promote this release: direct reaction at the surface of the iron oxide crystal; ligand-promoted exchange of iron ions; and the protonation of coordinating partners of iron ions to free the ions.

Inorganic phosphorus in soil is usually complexed with metal ions, a reaction that happens very fast. For example, the spike in concentration of soluble (bioavailable) phosphorus from addition of chemical fertilizers decreases to normal levels in a few days. This complexation is dependent on many factors, among them, the pH of the soil. Thus, the commonly accepted method for microbial-caused dissolution of phosphorus is through the activity of low molecular weight (i.e. highly soluble) organic acids. Though the exact mechanisms are still unknown, the accepted theory is that carboxylic acids act as chelating agents for the metal ions complexed with phosphorus, preventing precipitation of the free phosphorus ions released with a drop of pH. That is, the metal phosphorus complexes were disrupted by a drop in pH induced by the organic acids, the carboxylic acids caught the released metal ions, and therefore phosphorus ions was released into solution, where plants and microbes can uptake the phosphorus.

Volcanic Hot Springs

Sulfur deposits by microbes at a hot spring.

Thermophilic Microbes occur in volcanic hot springs around the world. Sulfur is an element commonly found in and around hot springs. The spring water is rich in sulfides that leach out of the rocks and soil, which is then oxidized by the microbes. The microbes strip electrons from the sulfide to use in their metabolic pathways, leaving behind sulfur. Other kinds of bacteria can oxidize sulfide or sulfur to sulfate, and form sulfuric acid. This is why many hot springs are acidic.

Different kinds of metabolic lifestyles can be found in different volcanic environments. For example, in the Kamchatka Hot Springs, lithotrophic methanogenesis was found to occur at alkaline conditions with a neutral pH. Sulfidogenesis occurred at neutral and acidic pH, and lithotrophic acetogenesis occurred over the entire range of pH. Lithotrophic reduction of Fe(III) also occurred over the entire range of pH. In these conditions, the only source of metabolic substrate comes from volcanic gases and methanogenesis is the process most commonly associated with hot spring microbial activity.

Examples of Applied Geomicrobiology

Geomicrobiology has many useful applications for humans. Many microbes are involved with breaking down toxic compounds, mineral extraction, and the manufacturing of valuable minerals and compounds.

Space Exploration

Geomicrobiology is being considered an important tool in the exploration and settlement of space because microorganisms can be used for a variety of uses, such as life support and waste management, energy production, extraction of industrially useful minerals, and the search for life using microbe-mineral interactions. Below are a couple of examples of how humans can use processes that are studied on earth to help explore and colonize space.

Life Support and Waste Management

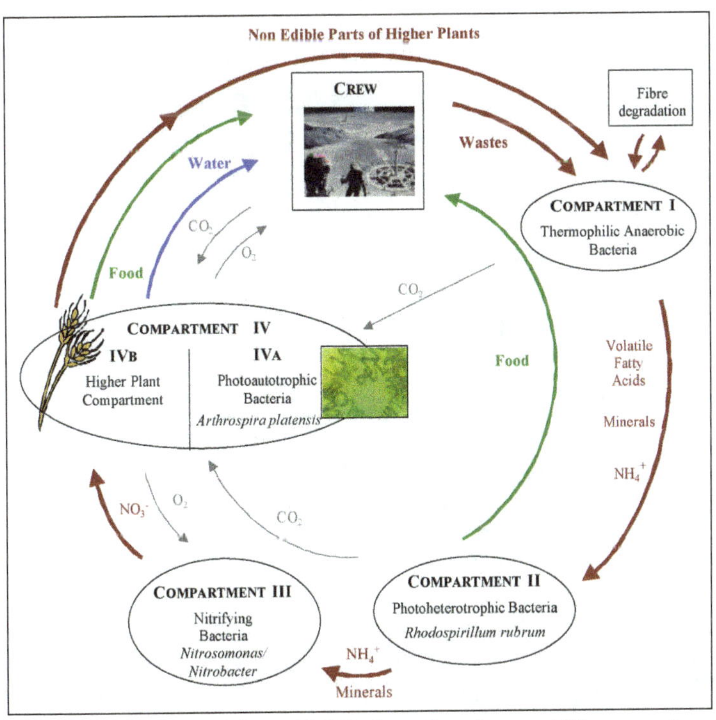

General concept of the MELISSA loop for advanced life support.

The biggest use of microbes in space exploration and settlement would be the use of photosynthetic organisms to convert water and carbon dioxide into oxygen and food in what is known as a biological life support system. Additionally the waste from these processes could be used to make more carbon dioxide, recycling the nutrients, water and carbon. The Micro Ecological Life Support System Alternative (MELiSSA), designed by European Space Agency, is a model system of an advance biological life support system that would be based on different species of microbes and plants. Consisting of many bioreactors and plant chambers, MELiSSA would break down the organic waste from the crew and non-edible parts of the higher plants. We can learn a lot about how MELiSSA works by looking at soil systems and processes here on Earth.

The MELiSSA loop is the basis of the life support system created by the ESA and uses processes we see all the time on earth. For example, the transformations of nitrogen (mineralization, nitrification, immobilization and denitrification) show the different ways nitrogen is transformed in soil and in the atmosphere. From studying how microbes interact in soil has helped us create MELiSSA, which combines multiple systems to produce food, water and oxygen for space explorers.

Energy Production

One example of the use of microbes in in the production of energy and additional waste management is the use of a MFC, or microbial fuel cell. MFCs can be used to generate electricity and process waste, but they currently produce low amounts of energy and depend on microbes processing organic matter in an anaerobic anode chamber. On earth MFCs can be made using the soil found all around us because soil contains a diverse population of microbes, along with the nutrients needed by those microbes to generate energy.

Microbial Mineral Resource Extraction

Gold Production, Detection and Transformation by Microbes

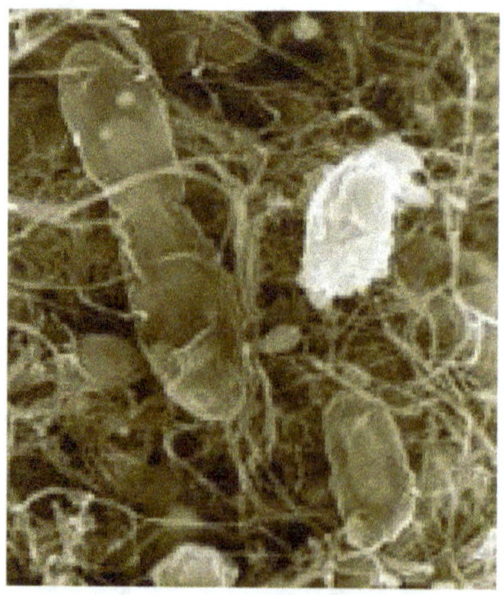

Cupriavidus metallidurans bacteria metabolize toxic gold chloride into gold nanoparticles (white).

Some bacteria are capable of reducing toxic gold chloride into pure gold. Cupriavidus metallidurans is one of the few bacteria that can survive in high concentrations of heavy metal in mines throughout the world. Although the bacteria only produce small amounts of gold nanoparticles too small to be profitable, they can be used by miners as biosensors to detect the concentrations of gold in environmental samples. Additionally, the gold nanoparticles synthesized by the bacteria could have potential in applications for many kinds of technology, such as optoelectronics, imaging technology, and drug delivery. Transcriptomic microarrays and DNA sequencing can detect bacterial genes that are associated with gold transformation, and thus pinpoint where significant gold deposits are present. There is great interest for the application of microbes in transforming toxic salts into gold, and scientists are studying both marine and terrestrial species to look for new insights into this area of geomicrobiology.

Role of Geomicrobiology in Bioleaching

Bioleaching can use microbes to extract metal from ore and remove metal impurities without the use of cyanide, mercury, or other chemicals. Biological forms of extraction can be cleaner than chemical methods because less toxins are released by the process, and detoxification can occur on site during processing. Chemolithotrophic bacteria reduce metal sulfides to metal sulfates, and heterotrophic bacteria or fungi exude organic acids, chelating, and complexing compounds to extract other ores and minerals. Although chemical extraction is still used when it is cheaper or another process is unknown, the field of bioleaching has high potential for further development.

The most active bacteria used in bioleaching are from genus Thiobacillus, gram negative bacteria that perform aerobic oxidation of sulfur compounds in acidic conditions. Other bioleaching genus' include iron- oxidizing Leptospirillum and Thermophilic bacteria.

Role of Geomicrobiology in Removal and Disposal of Heavy Metals and Pollutants

A plume from a mine waste spill in Durango, Colorado, with high Cromium (VI) content has spread to northwestern New Mexico, carried on the San Juan river.

Geomicrobiology can be applied to soil bioremediation, watershed reclamation, and to remove plumes of industrial contamination or radiation from the environment. Microbes in the soil and watertable are able to degrade contaminants that have no other practical source of removal and may be toxic to plants and other organisms. Inoculation of multiple target microbes can be used to remove multi-chemical pollutants from a site simultaneously. Selection and application of species must take into consideration the reactions that will take place at each step, to ensure the target compound is cleaned from the soil; for example, the desired reactions may require anaerobic or aerobic conditions in the soil and thus the introduced microbes must survive and thrive in such conditions.

Extraction processes bring heavy metals and other deep subsurface compounds to the surface soil or to the watershed resulting in contamination. Processing metal ore and other resources for production with heavy metals, like mercury, also contribute to the pollution. Microbes found within these layers have evolved to breakdown, reduce, or remove these compounds in their natural settings to extract energy and nutrients through use of precipitation, redox, sorption, and metabolic processes. Target microbes are inoculated at contaminated sites and the conditions are changed to select for the desired native microbial community that will transform the pollutant. Heavy metals including arsenic, zinc and sulfur can be precipitated out of acid mine wastewaters by sulfate-reducing bacteria, producing highly insoluble metal sulfides that are retained by wetland sediments.

Bacteria can reduce heavy metals from the soil in the rhizosphere for plant uptake and removal. Toxins can be converted to nontoxic forms and remain in the soil or be removed by translocation from the roots up into the plant, which can then be harvested to remove the metal from the site. For example, by providing anoxic conditions, high concentrations of toxic chromium (VI) are reduced by rhizospheric bacteria Arthrobacter and Pseudomonas to form immobile, nontoxic Cr(III). Chromium can be reduced from toxic Cr(VI) to nontoxic Cr(III) by endophytic bacteria within their cells; these Cr compounds are absorbed into the plant for easy removal from the soil and groundwater through harvesting. Some examples of endophytic bacteria species capable of Cr reduction include Ralstonia pickettii and Ochrobactrum intermedium (both gram -), and other non-endophytic examples of Cr-reducing bacteria include Bacillus cereus and Streptomyces (both gram +).

Role of Geomicrobiology in Radiation Contaminated Soils and Groundwater

Radiotrophic fungi can survive and colonize water and substrates (i.e. concrete barriers) in sites contaminated by severe radioactivity by using melanin pigment to transfer electrons and gain energy from the radioactive material. More study is needed to confirm if the radioactive material is then transformed to lower energy, safer products as the extra electrons are transferred by reduction. The fungal hyphae also deteriorate the concrete barriers by colonization and mineral weathering processes, which could result in release of radioactive material. A famous case is Chernobyl: after ten years of the catastrophe of Chernobyl in 1986 extensive fungal growth was observed on the walls and other building structures constructed in the inner part of the "Shelter" built over the fourth Unit of the Chernobyl nuclear power plant. Safe long-term storage of both existing and future nuclear waste is of vital importance in protecting the environment, therefore the study of these fungi has been key for future construction of nuclear waste repositories. Some of the microorganisms isolated from these extreme environments include modified strains from the genera Alternaria, Cladosporium, and Aureobasidium.

Role of Geomicrobiology in Agriculture

As agriculture enters the 21st century, new innovations are being discovered using microbes to increase production, yield, and sustainability in cropping systems. Soil quality and conservation are also becoming priorities on the agricultural forefront to combat the unintended consequences of conventional cropping systems. Metal and sulfate-reducing bacteria are responsible for precipitation and immobilization of metals like iron or manganese, and sulfate in the soil, determining the availability of these plant limiting nutrients.

For example, Phosphate solubilizing microorganisms (PSMs) use organic acids to dissociate phosphorus in small amounts from metal-phosphate compounds, making it available for plant uptake. PSM activity can result in reduced need for P fertilizer use and is being researched for use in biofertilizers. Some examples of PSMs include Chromobacterium, Azotobacter, and Bradyrhizobium. Rhizobia have been used for some time to help increase Nitrogen fixation in soils, however increasing knowledge of microbial control of nutrient availability are broadening sustainable approaches to improved crop and soil health and quality.

Microorganisms can be used in integrated pest management systems to fend off pests. The bacterium Thiobacillus, a genus of 'Hydrogenophilaceae', are thermophilic Proteobacteria growing in temperatures up to 50 °C. They obtain their energy from hydrogen oxidation and are used as pest control in potato fields to control scabs. Microbes are also used in less toxic insecticides called Microbials.

Magnetotactic Bacteria

Magnetotactic bacteria produce small magnetite crystals to form microbial compasses. These materials are used in building microchips and electronics.

For example, a number of magnetic nanomaterials are being synthesized with biomimetic techniques, which were inspired by the mineralization processes of iron these magnetotactic bacteria did, and are used in, to name a few, "high-density data storage and ferrofluidic devices, electromagnetic shielding, spintronics, and quantum computing, and applications in diagnostic medicine and targeted drug delivery".

Advances in Removal of Heavy Metals and Radioactivity

Research of microbial activities using radioactive and toxic metal waste in bedrock groundwater for safe long-term disposal is showing great potential for nuclear waste management. Corrosion of steel, copper, and other compounds is induced deep within the bedrock at greater rates than abiotic processes. The groundwater microbes bind the metal surfaces, forming biofilms resulting in corrosion of the metals.

Forensic Geology

Forensic geology is concerned with the application of geological data and techniques in relation to issues which may come before a court of law. It is closely related to environmental forensics, forensic engineering and forensic archaeology. Environmental forensics is somewhat broader in scope than forensic geology and involves a wider range of environmental data, knowledge and expertise. It frequently involves investigations of environmental problems such as water and air pollution. Forensic engineering also overlaps with environmental forensics and is typically concerned with such issues as ground stability, the failure of buildings and other engineering structures, flooding, wind damage, fires and explosions.

All sub-disciplines of the geosciences have potential forensic applications, but sedimentology, mineralogy, petrology, geochemistry, palaeontology and geophysics have so far made the greatest contributions. Shallow geophysical prospecting methods have been widely used by forensic archaeologists and others to locate and characterize clandestine graves and buried objects such as drugs and weapons. However, probably the most widely recognized application of forensic geology is the use of geological materials as trace evidence which can be of value in linking a suspect to a crime scene. In the wider forensic and legal literature, sediment, soil, dust and rock fragments have often been grouped together under the loose term 'soil' evidence.

Some of the earliest users of geological and soil evidence were in fact not geologists. As early as 1893 Hans Gross, an Austrian professor of criminology, had pointed out the value of examining 'dirt' on a suspect's shoes as a possible indicator of their movements, and the German chemist Georg Popp is widely credited as being the first to undertake systematic 'soil' comparison studies in the early 1900's. Around this time, the English fictional writer Sir Arthur Conan Doyle, literary creator of the investigator Sherlock Holmes, also utilized soil comparison in one of three cases in which he became a real life investigator. The work of these men had an important influence on Edmund Locard, initially a student of forensic medicine, who later went on to be Director of the Technical Police Laboratory in Lyons, France. Locard developed the first detailed scientific procedures for the analysis of dust traces, and established the famous Locard exchange principle:

Whenever two objects come into contact, there is always a transfer of material. The methods of detection may not be sensitive enough to demonstrate this, or the decay rate may be so rapid that all evidence of transfer has vanished after a given time. Nonetheless, the transfer has taken place.

In the United States, the Federal Bureau of Investigation Laboratory at Quantico, Virginia, was extensively using soil and mineral analyses in criminal cases as early as 1935, and has since maintained a strong interest in this area. Other important forensic work relating to particulates was undertaken in the 1960's and 1970's at McCrone Associates in Chicago, where the first Particle Atlas was developed. Outside the USA, scientists at the former UK Home Office Forensic Laboratory at Aldermaston, now closed, made extensive use of soil evidence in the 1960's and 1970's. In Japan, extensive use has been made of geological and soil evidence since the early 1980's by scientists at the National Research Institute of Police Science.

Since the early 1990's the true potential of geological and soil evidence, and that of other related sub-disciplines such as forensic botany, forensic entomology and forensic anthropology, has become much more widely recognized amongst police forces and forensic scientists worldwide. A much broader range of geological techniques and approaches is now being applied in the context of both criminal and civil law investigations. Many of these techniques also have extensive application in other areas, including the war against terrorism, international drug smuggling and broader environmental quality and public health campaigns. There is considerable current interest in applying forensic geological techniques and principles to such issues as illegal trading in ivory and rhino horn, archaeological artifacts and works of art, and in traceability studies related to a wide range of foodstuffs and other commodities.

Despite this long-standing interest, there is still only a relatively small specialist published literature in forensic geology, and until recently Murray and Tedrow's book 'Forensic Geology', originally published in 1975 and reprinted in 1992, was the only one available in the field.

Types of Information Provided by Geological and Soil Evidence

There are many questions and issues to which forensic geological techniques and information can be applied. These include:

- Was an individual, motor vehicle or other specified item present at a particular location?
- What was the sequence, and possible timing, of a visit to that location and possibly others?
- The location of buried objects (e.g. bodies, arms caches, drugs).
- The source of imported/smuggled items the cause of death (especially possible drowning and suffocation).
- The geographical origin of unidentified human remains.
- The length of time a body has been present at a location, and length of post-mortem interval.

Traces of rock, sediment, soil and dust can be present on a whole variety of items of interest, but amongst those most frequently submitted to the crime laboratory for examination are footwear, clothing, vehicles, flooring materials, digging implements, washing machine filters, polythene bags in which items have been stored, firearms and knives. Samples associated with the human body are also sometimes subject to examination. These include tapings of the skin, finger-nail scrapings, washings from the hair, nasal passages, trachea and lungs, contents of the gastro-intestinal tract and faeces.

With modern analytical techniques, only very small traces of mud on an exhibit can provide ample sample material for a whole battery of tests to be undertaken. However, the greater the amount of material which is available for analysis, the wider the options in terms of the analytical techniques which can be employed. If only small amounts of mud or particulates are present, there is usually a need to preserve as much of the evidence as possible for possible re-examination. Consequently nondestructive tests, or those which are minimally destructive, are preferable to destructive tests which require a relatively large amount of sample material.

Even exhibits which on first examination might seem unpromising from the viewpoint of preservation of forensic evidence, such as a burnt-out car, can in fact be a source of useful geological evidence. Many geological materials are not affected by temperatures typically found in standard vehicle fires, and in many instances particles of gravel and sand can be recovered from suitable parts of such vehicles, such as the footwells and suspension arms.

Persistence of Geological Evidence

Geological evidence may persist for a considerable period of time after it has been picked up from the source location. For example, gravel, sand or mud which enters the interior of a car on footwear, clothing or other items, such as a spade, will stay there, subject to some loss due to outward transfer on the footwear or clothing of later occupants, until such time as the vehicle is thoroughly cleaned. Even after cleaning, traces may remain in certain hard-to-access locations. Whereas some forms of forensic botanical evidence degenerate due to oxidation or fungal decomposition, most inorganic sediment particles are very resistant and may be immune to changes over time. Consequently, they provide useful clues years or even decades after an initial crime was committed, providing that exhibits have been retained and suitably stored to prevent environmental and cross-exhibit contamination.

Modification of Primary Transfer Soil Evidence

It should always be borne in mind that material initially picked up from a location, a process referred to as primary transfer, may subsequently be modified as additional particles are picked up from other locations, or as some of the primary particles are lost during subsequent movement. During primary transfer, the material transferred may not exactly reflect the exact nature of the material at the source point, depending on the nature of the material and the nature of the contact involved. For example, if a person sits or lies on wet ground there is a frequent tendency for the finer particles to be selectively transferred and retained on the clothing. In other circumstances, only a certain size range of coarse particles may be retained, for example coarser particles trapped within the detail of footwear sole treads, or gravel particles trapped in tyre treads. For this reason, it is always important that sample comparisons are undertaken on narrowly defined size fractions as well as on bulk samples.

Secondary Transfer

The possibility of secondary transfer of soil and other geological evidence should also always be borne in mind when exhibits are examined. For example, consider a case where Person A walks across a muddy car park and gets into the passenger seat of a vehicle and is driven by another person (B) to another location, where A gets out. During this process mud from the car park is transferred via the footwear of Person A to the front passenger footwell of the vehicle. The driver of the car (B) then drives to a third location and picks up another person, C, who also sits in the front passenger seat. The shoes of Person C come into contact with mud in the front passenger footwell deposited by Person A, and this is then transferred out of the car, via Peron C's footwear, onto the hallway carpet at Person C's home address. If it subsequently emerges that a crime has been committed at the car park, and Person C becomes a suspect, simple comparison of mud on Person C's footwear and hall carpet with control samples from the car park might lead to the spurious suggestion that Person C had been present at the crime scene. For this reason, great care needs to be taken by the forensic geologist to document the amount, distribution, layering and nature of any mud or similar evidence present on items, including footwear, submitted for analysis. In this connection there is an important responsibility on the part of police and scenes of crime examiners to provide the forensic geologist with necessary background information, and to ensure that comparison samples are taken from all locations and exhibits of possible relevance for comparative examination.

Geologic Modeling

Geologic modeling, or Geomodeling, is the applied science of creating computerized representations of portions of the Earth's crust based on geophysical and geological observations made on and below the Earth surface. A Geomodel is the numerical equivalent of a three-dimensional geological map complemented by a description of physical quantities in the domain of interest.

Geomodeling is related to the concept of Shared Earth Model; which is a multidisciplinary, interoperable and updatable knowledge base about the subsurface.

Geomodeling is commonly used for managing natural resources, identifying natural hazards, and quantifying geological processes, with main applications to oil and gas fields, groundwater aquifers and ore deposits. For example, in the oil and gas industry, realistic geologic models are required as input to reservoir simulator programs, which predict the behavior of the rocks under various hydrocarbon recovery scenarios. A reservoir can only be developed and produced once; therefore, making a mistake by selecting a site with poor conditions for development is tragic and wasteful. Using geological models and reservoir simulation allows reservoir engineers to identify which recovery options offer the safest and most economic, efficient, and effective development plan for a particular reservoir.

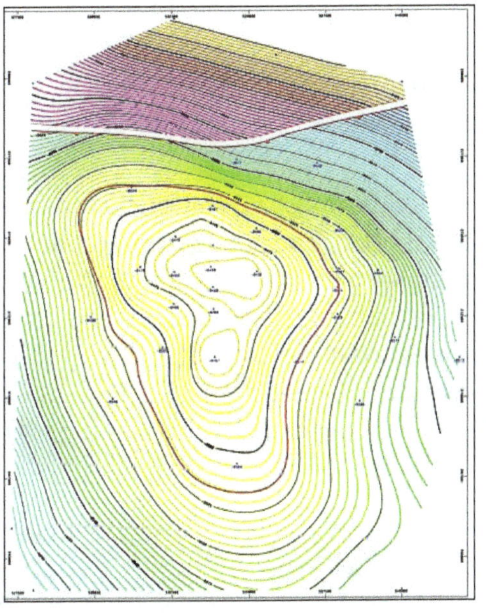

Geological mapping software displaying a screenshot of a structure map generated for an 8500ft deep gas & Oil reservoir in the Erath field, Vermilion Parish, Erath, Louisiana. The left-to-right gap, near the top of the contour map indicates a Fault line. This fault line is between the blue/green contour lines and the purple/red/yellow contour lines. The thin red circular contour line in the middle of the map indicates the top of the oil reservoir. Because gas floats above oil, the thin red contour line marks the gas/oil contact zone.

Geologic modelling is a relatively recent subdiscipline of geology which integrates structural geology, sedimentology, stratigraphy, paleoclimatology, and diagenesis.

In 2-dimensions (2D), a geologic formation or unit is represented by a polygon, which can be bounded by faults, unconformities or by its lateral extent, or crop. In geological models a geological unit is bounded by 3-dimensional (3D) triangulated or gridded surfaces. The equivalent to the mapped polygon is the fully enclosed geological unit, using a triangulated mesh. For the purpose of property or fluid modelling these volumes can be separated further into an array of cells, often referred to as voxels (volumetric elements). These 3D grids are the equivalent to 2D grids used to express properties of single surfaces.

Geomodelling generally involves the following steps:

1. Preliminary analysis of geological context of the domain of study.
2. Interpretation of available data and observations as point sets or polygonal lines (e.g. "fault sticks" corresponding to faults on a vertical seismic section).
3. Construction of a structural model describing the main rock boundaries (horizons, unconformities, intrusions, faults).
4. Definition of a three-dimensional mesh honoring the structural model to support volumetric representation of heterogeneity and solving the Partial Differential Equations which govern physical processes in the subsurface (e.g. seismic wave propagation, fluid transport in porous media).

Geologic Modeling Components

Structural Framework

Incorporating the spatial positions of the major formation boundaries, including the effects of faulting, folding, and erosion (unconformities). The major stratigraphic divisions are further subdivided into layers of cells with differing geometries with relation to the bounding surfaces (parallel to top, parallel to base, proportional). Maximum cell dimensions are dictated by the minimum sizes of the features to be resolved (everyday example: On a digital map of a city, the location of a city park might be adequately resolved by one big green pixel, but to define the locations of the basketball court, the baseball field, and the pool, much smaller pixels – higher resolution – need to be used).

Rock Type

Each cell in the model is assigned a rock type. In a coastal clastic environment, these might be beach sand, high water energy marine upper shoreface sand, intermediate water energy marine lower shoreface sand, and deeper low energy marine silt and shale. The distribution of these rock types within the model is controlled by several methods, including map boundary polygons, rock type probability maps, or statistically emplaced based on sufficiently closely spaced well data.

Reservoir Quality

Reservoir quality parameters almost always include porosity and permeability, but may include measures of clay content, cementation factors, and other factors that affect the storage and deliverability of fluids contained in the pores of those rocks. Geostatistical techniques are most often used to populate the cells with porosity and permeability values that are appropriate for the rock type of each cell.

Fluid Saturation

Most rock is completely saturated with groundwater. Sometimes, under the right conditions, some of the pore space in the rock is occupied by other liquids or gases. In the energy industry, oil and natural gas are the fluids most commonly being modelled. The preferred methods for calculating hydrocarbon saturations in a geologic model incorporate an estimate of pore throat size, the densities of the fluids, and the height of the cell above the water contact, since these factors exert the strongest influence on capillary action, which ultimately controls fluid saturations.

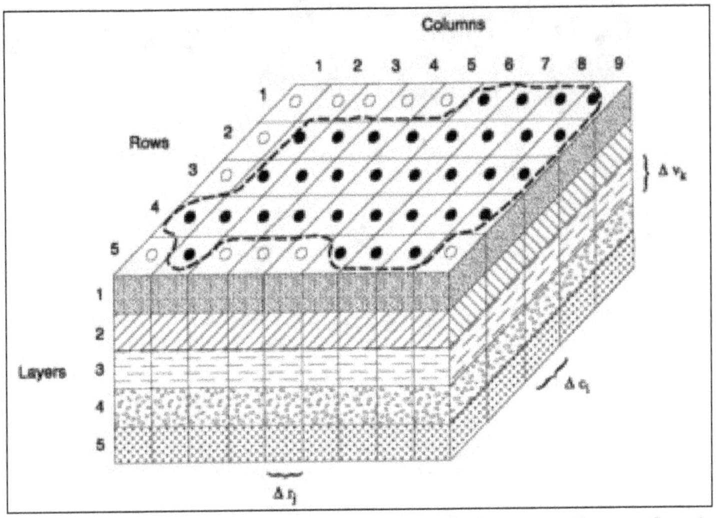

A 3D finite difference grid used in MODFLOW for simulating groundwater flow in an aquifer.

Geostatistics

An important part of geologic modelling is related to geostatistics. In order to represent the observed data, often not on regular grids, we have to use certain interpolation techniques. The most widely used technique is kriging which uses the spatial correlation among data and intends to construct the interpolation via semi-variograms. To reproduce more realistic spatial variability and help assess spatial uncertainty between data, geostatistical simulation based on variograms, training images, or parametric geological objects is often used.

Mineral Deposits

Geologists involved in mining and mineral exploration use geologic modelling to determine the geometry and placement of mineral deposits in the subsurface of the earth. Geologic models help define the volume and concentration of minerals, to which economic constraints are applied to determine the economic value of the mineralization. Mineral deposits that are deemed to be economic may be developed into a mine.

Technology

Geomodelling and CAD share a lot of common technologies. Software is usually implemented using object-oriented programming technologies in C++, Java or C# on one or multiple computer platforms. The graphical user interface generally consists of one or several 3D and 2D graphics windows to visualize spatial data, interpretations and modelling output. Such visualization is generally achieved by exploiting graphics hardware. User interaction is mostly performed through mouse and keyboard, although 3D pointing devices and immersive environments may be used in some specific cases. GIS (Geographic Information System) is also a widely used tool to manipulate geological data. Geometric objects are represented with parametric curves and surfaces or discrete models such as polygonal meshes.

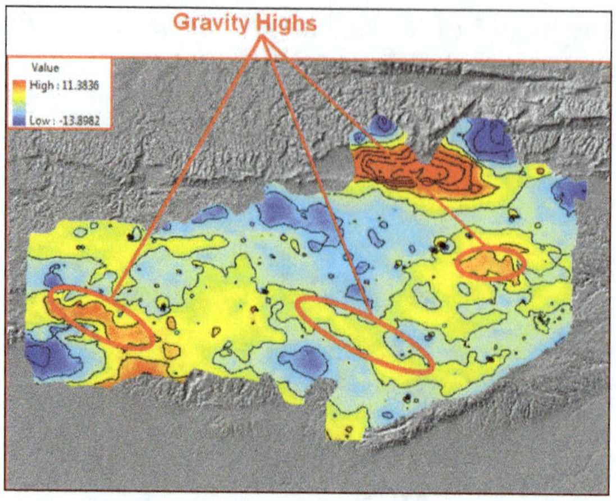

Gravity Highs over the Mardin Uplift.

Geomodeling

Problems pertainting to Geomodelling cover:

- Defining an appropriate Ontology to describe geological objects at various scales of interest.
- Integrating diverse types of observations into 3D geomodels: geological mapping data, borehole data and interpretations, seismic images and interpretations, potential field data, well test data, etc.
- Better accounting for geological processes during model building.
- Characterizing uncertainty about the geomodels to help assess risk. Therefore, Geomodelling has a close connection to Geostatistics and Inverse problem theory.
- Applying of the recent developed Multiple Point Geostatistical Simulations (MPS) for integrating different data sources.
- Automated geometry optimization and topology conservation.

Geologic Modeling Software

Software developers have built several packages for geologic modelling purposes. Such software can display, edit, digitise and automatically calculate the parameters required by engineers, geologists and surveyors. Current software is mainly developed and commercialized by oil and gas or mining industry software vendors:

- Geologic modeling and visualisation:
  - SGS Genesis,
  - IRAP RMS Suite,
  - Geomodeller3D,
  - Geosoft provides GM-SYS and VOXI 3D modelling software,

   - GSI3D,
   - Petrel,
   - Rockworks,
   - Move,
- Groundwater modeling:
   - FEFLOW,
   - FEHM,
   - MODFLOW,
   - GMS,
   - Visual MODFLOW,
   - ZOOMQ3D.

Industry Consortia or companies are specifically working at improving standardization and interoperability of earth science databases and geomodelling software:

- Standardization: GeoSciML by the Commission for the Management and Application of Geoscience Information, of the International Union of Geological Sciences.
- Standardization: RESQML(tm) by Energistics.
- Interoperability: OpenSpirit, by TIBCO(r).

References

- Economic-geology-and-environment- 42160: yourarticlelibrary.com, Retrieved 19 February, 2019
- Planetary-geology, encyclopedias-almanacs-transcripts-and-maps: encyclopedia.com, Retrieved 23 August, 2019
- Geomicrobiology: microbewiki.kenyon.edu, Retrieved 25 April, 2019
- Forensicgeologyencyclopedia: course1.winona.edu, Retrieved 25 February, 2019
- Geologic-modelling: courses.lumenlearning.com, Retrieved 16 June, 2019

PERMISSIONS

All chapters in this book are published with permission under the Creative Commons Attribution Share Alike License or equivalent. Every chapter published in this book has been scrutinized by our experts. Their significance has been extensively debated. The topics covered herein carry significant information for a comprehensive understanding. They may even be implemented as practical applications or may be referred to as a beginning point for further studies.

We would like to thank the editorial team for lending their expertise to make the book truly unique. They have played a crucial role in the development of this book. Without their invaluable contributions this book wouldn't have been possible. They have made vital efforts to compile up to date information on the varied aspects of this subject to make this book a valuable addition to the collection of many professionals and students.

This book was conceptualized with the vision of imparting up-to-date and integrated information in this field. To ensure the same, a matchless editorial board was set up. Every individual on the board went through rigorous rounds of assessment to prove their worth. After which they invested a large part of their time researching and compiling the most relevant data for our readers.

The editorial board has been involved in producing this book since its inception. They have spent rigorous hours researching and exploring the diverse topics which have resulted in the successful publishing of this book. They have passed on their knowledge of decades through this book. To expedite this challenging task, the publisher supported the team at every step. A small team of assistant editors was also appointed to further simplify the editing procedure and attain best results for the readers.

Apart from the editorial board, the designing team has also invested a significant amount of their time in understanding the subject and creating the most relevant covers. They scrutinized every image to scout for the most suitable representation of the subject and create an appropriate cover for the book.

The publishing team has been an ardent support to the editorial, designing and production team. Their endless efforts to recruit the best for this project, has resulted in the accomplishment of this book. They are a veteran in the field of academics and their pool of knowledge is as vast as their experience in printing. Their expertise and guidance has proved useful at every step. Their uncompromising quality standards have made this book an exceptional effort. Their encouragement from time to time has been an inspiration for everyone.

The publisher and the editorial board hope that this book will prove to be a valuable piece of knowledge for students, practitioners and scholars across the globe.

INDEX

www.ingramcontent.com/pod-product-compliance
Lightning Source LLC
LaVergne TN
LVHW080921210926
848485LV00004B/701